K.W. Bagnall, *The Actinide Elements,* in P.L. Robinson (Ed.), *Topics in Inorganic and General Chemistry,* Monograph 15, Elsevier Publishing Company, Amsterdam, 1972.

ERRATA

page 18, line 4, should read:

and neodymium, can be oxidised to this higher oxidation state and all of

page 171, 2nd par., line 4, should read:

about $100°$. A uranyl(VI) tellurate of composition $UO_2 TeO_4 \cdot 2 Na_2 TeO_4 \cdot 4 H_2 O$

THE ACTINIDE ELEMENTS

TOPICS IN INORGANIC
AND GENERAL CHEMISTRY

A COLLECTION OF MONOGRAPHS EDITED BY

P.L. ROBINSON

Emeritus Professor of Chemistry in the University of Durham and the University of Newcastle upon Tyne

MONOGRAPH 15

Other titles in the collection:

In preparation

THE ACTINIDE ELEMENTS

BY

K.W. BAGNALL, D.Sc., F.R.I.C.,

Professor of Inorganic Chemistry,
The University of Manchester (Great Britain)

ELSEVIER PUBLISHING COMPANY
AMSTERDAM / LONDON / NEW YORK
1972

ELSEVIER PUBLISHING COMPANY
335 JAN VAN GALENSTRAAT
P.O. BOX 211, AMSTERDAM, THE NETHERLANDS

AMERICAN ELSEVIER PUBLISHING COMPANY, INC.
52 VANDERBILT AVENUE
NEW YORK, N.Y. 10017

LIBRARY OF CONGRESS CARD NUMBER: 72-83194

ISBN 0-444-41041-4

WITH 35 ILLUSTRATIONS AND 35 TABLES

PRINTED IN THE NETHERLANDS

Preface

The discovery of the first synthetic transuranium elements, neptunium and plutonium, and the need to develop the production of the latter for military purposes from 1940 onwards (the Manhattan project) stimulated research on the chemistry of the new elements as well as on the synthesis of elements of atomic number higher than that of plutonium. The latter resulted in the completion of the actinide series in 1961 with the successful synthesis of lawrencium. The interest in research on the chemistry of the actinides grew steadily from 1940, reaching a peak in the middle 1960's. During that time a great deal of the basic chemistry of the members of the group up to americium was reported, and more recently a considerable extension has been made to our knowledge of the elements beyond americium as these have become more readily available, following their large-scale production in the United States of America.

The results of the earlier research, together with the new information on the higher actinides mentioned above, have been used to provide what the writer believes to be a balanced account of the chemistry of the actinides within the covers of a short monograph. In a number of instances the chemical behaviour of the actinides is correlated with that of the lanthanides, showing up similarities and differences between the two series of f transition elements. The resulting account is comprehensive, not in the sense of containing every detail, but as a broad survey in some depth which should be useful to those studying or teaching the subject at honours level. This survey has, however, a sufficiently strong backing of references for it to also make accceptable reading to those already engaged in research in this interesting field and may also serve as a helpful point of departure for those planning to undertake new work in the field of actinide chemistry.

Much of the information now presented has not previously been collected together elsewhere. The writer has had to be selective in the choice of the material for presentation in order to restrict the monograph to manageable proportions. To give an instance, complexes of the actinide halides are discussed only by reference to the common stoicheiometries rather than by being given full coverage, for to do this would burden the reader with a considerable mass of almost identical detail. Reviews of this area of actinide chemistry are, however, available and reference is made to them. Other, less well known topics, such as, for example, the alkoxides, carboxylates, chelate complexes and organometallic compounds are treated in considerable detail; in these instances the coverage is as complete as the writer can make it. The monograph closes with a short chapter on f orbitals, and the magnetic properties and spectra of the actinides, subjects to which reference has been made in a number of places in the monograph and which are not commonly dealt with at all in the standard textbooks of inorganic chemistry.

I am greatly indebted to Professor P.L. Robinson for critically reading the drafts of the book and for the helpful suggestions which he made at each stage. I am also grateful to a number of my colleagues who have read various parts of the manuscript and suggested the many improvements which have been incorporated into this final version. Finally, I wish to thank the following for permission to reproduce illustrations from their journals; the American Chemical Society, the Chemical Society and the publishers of Acta Chemica Scandinavica, Acta Crystallographica, the Journal of Inorganic and Nuclear Chemistry, Molecular Physics, Nature and Science Progress, as well as the publishers of two books, namely *The Chemistry of the Actinide Elements* by J.J. Katz and G.T. Seaborg and *Man-Made Transuranium Elements* by G.T. Seaborg.

Bramhall, Cheshire,   K.W. Bagnall
December 1971

Contents

Chapter 1

THE DISCOVERY AND OCCURRENCE OR SYNTHESIS OF THE ACTINIDES

1. INTRODUCTION

The actinide series comprises the fourteen elements following actinium ($Z = 89$) and is analogous to the lanthanide, or rare earth, series in that both result from the filling of the inner $4f$ and $5f$ shells respectively. Although actinium itself is not strictly an actinide element, it is included in the discussion for the sake of comparisons to be made with the chemistry of later elements in the group.

Until 1940, only the first four elements, actinium, thorium, protactinium and uranium, were known; all of them are radioactive, as would be expected from their position at the end of the Periodic Table as it was then known, and it was not until the first of the transuranium elements had been synthesised that the analogy between these elements and the lanthanides was recognised. The classification as a second series of f transition elements was due to G.T. Seaborg, who was personally involved in the discovery and identification of nearly all the transuranium elements.

The filling of the f shell across the lanthanide and actinide groups does not mean that these elements all have the same nd^1 $(n + 1)s^2$ valence configuration; the actual ground state electron configurations of the elements of the two groups are given in Table 1. However, from this it can be seen that, from plutonium onwards, the actinides generally have the same electronic configurations as their lanthanide analogues, whereas the earlier members of the actinide group retain some d electrons in their ground state configurations. Initially in the actinide series the $5f$ level is higher in energy than the $6d$, but as the atomic number increases the energy of the $5f$ level decreases below that of the $6d$ (Fig. 1).

TABLE 1

ELECTRONIC CONFIGURATIONS OF THE $4f$ AND $5f$ ELEMENTS

(*Xe core*)			(*Rn core*)		
Lanthanum	(La)	$5d6s^2$	Actinium	(Ac)	$6d7s^2$
Cerium	(Ce)	$4f^2 6s^2$	Thorium	(Th)	$6d^2 7s^2$
Praseodymium	(Pr)	$4f^3 6s^2$	Protactinium	(Pa)	$5f^2 6d7s^2$ or $5f6d^2 7s^2$
Neodymium	(Nd)	$4f^4 6s^2$	Uranium	(U)	$5f^3 6d7s^2$
Promethium	(Pm)	$4f^5 6s^2$	Neptunium	(Np)	$5f^4 6d7s^2$ or $5f^5 7s^2$
Samarium	(Sm)	$4f^6 6s^2$	Plutonium	(Pu)	$5f^6 7s^2$
Europium	(Eu)	$4f^7 6s^2$	Americium	(Am)	$5f^7 7s^2$
Gadolinium	(Gd)	$4f^7 5d6s^2$	Curium	(Cm)	$5f^7 6d7s^2$
Terbium	(Tb)	$4f^9 6s^2$	Berkelium	(Bk)	$5f^9 7s^2$ or $5f^8 6d7s^2$
Dysprosium	(Dy)	$4f^{10} 6s^2$	Californium	(Cf)	$5f^{10} 7s^2$
Holmium	(Ho)	$4f^{11} 6s^2$	Einsteinium	(Es)	$5f^{11} 7s^2$
Erbium	(Er)	$4f^{12} 6s^2$	Fermium	(Fm)	$5f^{12} 7s^2$
Thulium	(Tm)	$4f^{13} 6s^2$	Mendelevium	(Md)	$5f^{13} 7s^2$
Ytterbium	(Yb)	$4f^{14} 6s^2$	Nobelium	(No)	$5f^{14} 7s^2$
Lutetium	(Lu)	$4f^{14} 5d6s^2$	Lawrencium	(Lw)	$5f^{14} 6d7s^2$

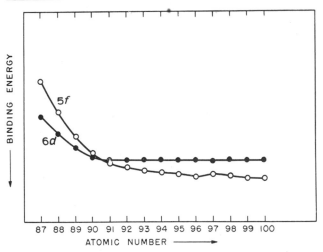

Fig. 1. A qualitative representation of the electronic binding energies in the heaviest elements (J.J. Katz and G.T. Seaborg, *The Chemistry of the Actinide Elements*, Methuen, London, 1957, p. 465. Reproduced by permission of the publishers.

2. THE DISCOVERY OF ELEMENTS 89-92

(a) Uranium

Historically uranium is the longest known of all the actinide elements; it was discovered in pitchblende, then thought to be an iron tungstate, by M.H. Klaproth in 1789 and named after Uranus, the planet discovered by Herschel in 1781. Until 1872, when D. Mendeleev assigned uranium to its present position in the Periodic Table, giving it an approximate atomic weight of 240, uranium was thought to have an atomic weight of only 120 which had frustrated earlier attempts to fit it into periodic classifications.

Uranium is by no means rare, occurring[1] to the extent of about 4 ppm in the outermost layer of the earth's crust, its abundance being appreciably greater than that of antimony, bismuth, cadmium, iodine, mercury or silver. The principal minerals are oxides, such as uraninite (UO_2), pitchblende $(UO_{2.2}-UO_{2.67})$ and uranyl vanadates, such as carnotite $[K_2(UO_2)_2(VO_4)_2 \cdot 1-3H_2O]$ or phosphates, such as autunite $[Ca(UO_2)_2(PO_4)_2 \cdot 8-12H_2O]$, this last being the most common uranium mineral[2].

Towards the end of the nineteenth century, H. Becquerel observed the phenomenon of radioactivity in uranium, thus stimulating research which led to the discovery of a number of other radioelements in this region of the Periodic Table, and, eventually, to the development of the concepts of the nuclear atom and of isotopes. The two principal isotopes present in natural uranium are ^{235}U (0.72 atoms %), which is fissionable by thermal, or slow, neutrons and ^{238}U (99.2%) which is not, and the element became important industrially as a source of nuclear power after the discovery of fission by O. Hahn and F. Strassmann in 1938/39. Before that time almost the only industrial application of uranium was as a colouring material in the manufacture of yellow glass; examples of uranium coloured glass are known which date back to A.D. 79.

(*b*) *Thorium*

Over thirty years after the discovery of uranium, J.J. Berzelius took up again his earlier (1815) work on an oxide which he had named thorine, after the mythological Scandinavian god of war. He had originally thought that the oxide resembled zirconium dioxide, but in 1824 he showed that the parent mineral consisted mainly of yttrium phosphate. In 1828, however, he found that another mineral from Lövö in Norway did indeed contain a new element, chemically analogous to zirconium, to which he gave the name thorium. The mineral is now known as thorite.

Thorium is more abundant than uranium in the earth's crust, the average content being 12 ppm[1], close to that of lead (16 ppm). The principal minerals[2,3] are the silicates thorite and huttonite ($ThSiO_4$) and a hydroxo-silicate variant of thorite, thorogummite, $Th(SiO_4)_{1-x}$-$(OH)_{4x}$, in which tetrahedral groups of four hydroxide ions randomly replace the silicate ions in the lattice. The most important source of thorium is, however, monazite; this is a mixture of lanthanide and thorium phosphates which can contain from 10 to 30% of thorium dioxide.

Following the discovery of the radioactivity of uranium, other heavy elements were investigated and in 1898 Mme. M.S. Curie and G.C. Schmidt independently showed that thorium was also radioactive. Natural thorium is almost entirely ^{232}Th; although this isotope is not fissionable by thermal neutrons it undergoes thermal neutron capture to yield ^{233}U, a fairly long-lived isotope of uranium (1.6×10^5 yr) which does not occur in nature.

$$^{232}_{90}Th(n, \gamma) \quad ^{233}_{90}Th \quad \xrightarrow[22.4 \text{ min}]{\beta} \quad ^{233}_{91}Pa \quad \xrightarrow[27.0 \text{ days}]{\beta} \quad ^{233}_{92}U$$

This isotope of uranium is fissionable by neutrons and suitable as a nuclear reactor fuel. Indeed its formation is important as the basis of the "breeder" reactor, in which escaping thermal neutrons in the outerlying areas of the reactor are captured by thorium instead of being

wasted. In this way more nuclear fuel is produced than is consumed in
the reactor.

Thorium itself remained industrially unimportant until C.A. von
Welsbach's investigations of the emission of light by strongly heated
lanthanide oxides (1880–1890), which led ultimately to the discovery
that thorium dioxide containing about 1% cerium gave a much greater
light emission than any other oxide when heated, a discovery which led
to the widespread use of thorium in incandescent gas-mantles. With the
decline of gas as a means of illumination, the demand for thorium
dioxide decreased; its chief use then was as a refractory material for the
manufacture of special crucibles used in high temperature work.
Nowadays a principal application is for "breeder" reactors, as mentioned
above.

The remaining two naturally occurring elements in this group,
actinium and protactinium, have no isotopes of sufficiently long half-
life to exist independently in nature, but their longest-lived isotopes,
^{227}Ac (22 yr) and ^{231}Pa (32 500 yr) are both members of the ^{235}U
decay chain:

$$^{235}_{92}U \xrightarrow[7.1 \times 10^8 \text{ yr}]{\alpha} {}^{231}_{90}Th \xrightarrow[25.6 \text{ h}]{\beta} {}^{231}_{91}Pa \xrightarrow{\alpha} {}^{227}_{89}Ac$$

(c) Actinium

The discovery of actinium is usually attributed to Debierne, who
claimed to have isolated it in 1899, but the actual discovery of the
element is probably due[8] to Giesel (1900–1902). The name is derived
from the Greek ἄκτις, ἄκτινος, a beam or ray, and refers to its radioactivity.
Since ^{227}Ac has an extremely short half-life compared with that of
^{235}U, the quantity of the element in equilibrium with one ton of
elementary natural uranium is only 0.2 mg and, since the chemical
behaviour of actinium is very similar to that of the lanthanides which
are generally also present in uranium ores, the separation of the element
in a pure state from uranium ores is extremely difficult. There are,
however, two alternatives; the parent ^{231}Pa is more easily separable

from the ores, and although the growth of actinium into the separated ^{231}Pa would be a slow process, it would be relatively easy to separate the actinium from its parent. The other alternative is to synthesise ^{227}Ac by neutron irradiation of ^{226}Ra:

$$^{226}_{88}Ra\,(n,\,\gamma)\quad ^{227}_{88}Ra\quad \xrightarrow[41\ min]{\beta}\quad ^{227}_{89}Ac$$

This method is quite successful[4] and most of the known chemistry of actinium has been investigated with milligram amounts of the element obtained in this way.

(d) Protactinium

A short-lived isotope of protactinium, ^{234}Pa, a product of the ^{238}U decay chain, was discovered by K. Fajans and O. Göhring in 1913 and named "brevium" on account of the shortness of its half-life, 6.66 h and 1.2 min respectively for the two nuclear isomers, so clearly ^{234}Pa is useless for chemical studies. The longer-lived isotope, ^{231}Pa (32 500 yr) was discovered by O. Hahn and L. Meitner in 1917 and, independently, by F. Soddy and J.A. Cranston at about the same time; it was this isotope of the element which provided the name protactinium, implying that it is the parent of actinium. The equilibrium concentration of ^{231}Pa in natural uranium is only about 300 mg per ton of elementary uranium, and apart from the pioneering work of A. von Grosse, which culminated in the separation of about 100 mg of the element in the late 1930s, little was known about the chemistry of protactinium until 1960, when over 100 g of the element was separated from uranium refinery residues at the U.K.A.E.A. Springfields and Windscale plants, in collaboration with a group at A.E.R.E., Harwell. The historical aspects of this separation have been described in a review[5].

3. THE DISCOVERY OF ELEMENTS 93–100

Following the discovery of the neutron by J. Chadwick in 1932, a

number of attemps were made to synthesise transuranium elements by
bombarding uranium with neutrons, but chemical studies of the products
of these reactions showed that thermal-neutron fission had occurred and
it was not until 1940 that the formation of an isotope of element 93,
neptunium, named after the planet beyond Uranus, was established by
E.M. McMillan. He, with P.H. Abelson, showed that, in addition to
neutron-induced fission, a short-lived (2.35 day) β-emitting isotope of
element 93 was formed by thermal-neutron capture by ^{238}U:

$$^{238}_{92}U\,(n,\gamma)\quad ^{239}_{92}U\quad \xrightarrow[\text{23.5 min}]{\beta}\quad ^{239}_{93}Np$$

The formation of elements 93–100 by neutron irradiation is shown in
Fig. 2.

The long-lived (2.2×10^6 yr), α-emitting nuclide, ^{237}Np is much
more useful than ^{239}Np for chemical studies on the element and this is
formed in two different ways in a nuclear reactor. ^{235}U, as well as
undergoing fission by thermal neutrons, can also capture neutrons and
two successive neutron captures lead to the formation of ^{237}U, which
decays by β-emission to ^{237}Np. At the same time, the more plentiful
^{238}U can undergo $(n, 2n)$ reactions with fast neutrons, as before
yielding ^{237}U (Fig. 2).

Although ^{239}Np decays by β-emission, the α-emitting decay
product, ^{239}Pu, was not discovered until 1941 and the first report of
element 94 arose from the work of E.M. McMillan, J.W. Kennedy, A.C.
Wahl and G.T. Seaborg who showed, at the end of 1940, that cyclotron
bombardment of ^{238}U with deuterons yielded ^{238}Np which decays to
the 86.4 yr α-emitting ^{238}Pu:

$$^{238}_{92}U + {}^{2}_{1}H \rightarrow 2\,{}^{1}_{0}n + {}^{238}_{93}Np \quad \xrightarrow[2.10\,d]{\beta} \quad {}^{238}_{94}Pu$$

As with neptunium, the name is derived from a planet, in this instance,

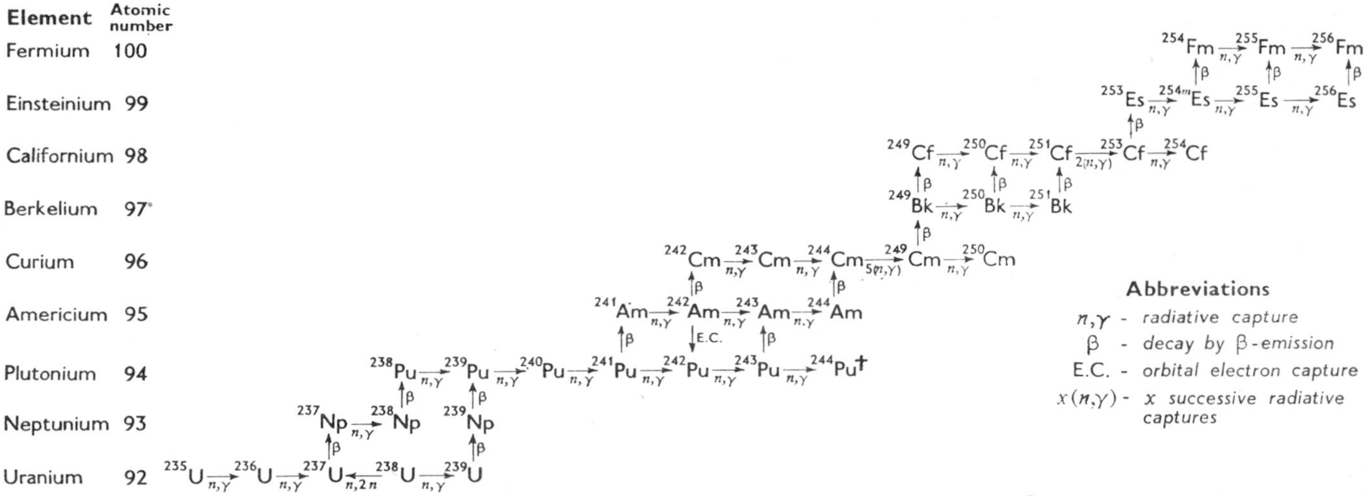

† α-emitter of $\sim 7{\cdot}6 \times 10^7$ year half life; neutron capture gives rise to ^{245}Pu and ^{246}Pu, both of which decay by β-emission to ^{245}Am and ^{246}Am, and then to ^{245}Cm and ^{246}Cm

Fig. 2. The formation of the transuranium elements in a nuclear reactor (K.W. Bagnall, *Sci. Progr.*, 52 (1964) 68). Reproduced by permission of the publishers, Blackwell, Oxford.

Pluto, the last discovered planet, which lies beyond Uranus.

Although neither neptunium-237 nor plutonium-239 can exist on their own in nature because their half-lives are relatively short (2.2×10^6 yr and 24 300 yr respectively) compared with the age of the earth ($\sim 4 \times 10^9$ yr), traces of them, amounting to about 1 g in 10^6 tons of uranium, have been detected in uranium minerals. These small amounts represent equilibrium concentrations of these two isotopes arising from neutron bombardment of ^{238}U in the ores, as in their production in a nuclear reactor. The neutrons required for their production in the ore come from α, n reactions with light elements, such as oxygen or silicon, in the ores and, to some extent, from the spontaneous fission of the uranium present.

Americium and curium, named respectively after the continent of America, by analogy with europium, and after Pierre and Marie Curie, by analogy with gadolinium (J. Gadolin, the Finnish chemist) were not made until 1944/45, G.T. Seaborg being actively involved in the synthesis of these and the remaining actinide elements. The making of curium preceded that of americium, the first synthesis being that of ^{242}Cm by helium ion bombardment of ^{239}Pu:

$$^{239}_{94}Pu + {}^{4}_{2}He \rightarrow {}^{242}_{96}Cm + {}^{1}_{0}n$$

This isotope is now more conveniently made by neutron irradiation of ^{241}Am (Fig. 2). Soon afterwards it was found that ^{241}Pu, formed from ^{239}Pu by two successive neutron captures, decayed by β-emission to ^{241}Am. These nuclides were then used as target materials in the work leading to the discovery of elements 97 and 98, which was made in late 1949 and early 1950. These two elements were named respectively berkelium and californium after the city of Berkeley, were the discovery was made in the University of California laboratories, and after the state of California where Berkeley is situated. The naming of element 97 follows the lanthanide analogue in that terbium derives from Ytterby in Sweden, but there is no such analogy in the case of element 98 for the name of corresponding lanthanide, dysprosium, is derived from the Greek "difficult to get at".

Elements 99 and 100, einsteinium and fermium, named after the physicists A. Einstein and E. Fermi, were discovered in the fall-out debris from the "Mike" thermonuclear explosion test of November 1, 1952; they had obviously been built up by a multiple neutron capture process, which is now thought to occur in two stages. The first is a fast neutron reaction with ^{238}U, yielding ^{238}Pa:

$$^{238}_{92}U + {}^{1}_{0}n \rightarrow {}^{238}_{91}Pa + {}^{1}_{1}H$$

This nuclide then undergoes multiple neutron capture to yield a range of very heavy, β-unstable protactinium isotopes in the mass range 244–257 which decay to heavy actinide nuclides which include $^{253}_{99}$Es, $^{255}_{99}$Es, $^{255}_{100}$Fm and $^{257}_{100}$Fm. All of these heavier actinides are now made by intense and prolonged neutron bombardment of ^{239}Pu in nuclear reactors (Fig. 2).

All of the neutron capture reactions shown in Fig. 2 refer to thermal neutrons; since the probability of thermal neutron fission is, in many cases, as high as, or higher than, the probability of neutron capture for nuclides which are made up either of odd numbers of protons and odd numbers of neutrons (odd-odd nuclides) or of even numbers of protons and odd numbers of neutrons (even-odd nuclides), intermediate nuclides in the capture chain are liable to be destroyed by fission to a considerable extent.

The probabilities of fission and capture are expressed in terms of the effective cross sections, σ, of the nuclides concerned and these are expressed as an area, the unit being the barn ("as big as a barn" being the apocryphal comment concerning a nuclide observed to have a high fission probability). The barn is equal to 10^{-24} cm^2.

Radioactive decay data for the transuranium actinide nuclei show that the α-decay half-lives for elements beyond curium decrease quite dramatically, and a competitive mode of decay becomes important in this region, decay by spontaneous fission. The combination of loss of product or intermediate nuclide by α- or spontaneous fission decay and by thermal-neutron fission means that the yields of the higher actinides become progressively smaller as the atomic number increases and there is no evidence at all for the formation of elements 101–103 in nuclear reactors. Because of these ever-decreasing yields of the elements up to fermium, the possibility of synthesising them by using the very intense, but short-lived, neutron flux generated on the explosion of an atomic bomb has been investigated (*cf.* the discovery of einsteinium and fermium mentioned earlier). The procedure is to bury the bomb in an underground cavern in order to prevent the escape to the atmosphere of the products of the explosion, the problem then hingeing on the difficulty of working up very small amounts of the product nuclei from the many tons of compacted and fused rock resulting from the high temperature and pressure generated by the explosions. The tests carried out so far seem to indicate the limit to be ^{257}Fm for this type of reaction.

Apart from the use of ^{239}Pu as a reactor fuel or in atomic bombs, the only other application of these elements depends on their radioactivity, in which the heat produced by the α-emitting isotopes is used. This heat arises from the deceleration of the disintegration α-particles within the sample of radioactive material and its container. Because of the heat emission, many of the actinide nuclides are very useful for light-weight heat sources for the generation of thermo-electric power in space satellites, heart pace-makers and other devices, the basis of their operation being a direct conversion of heat into electricity by means of the principle of the conventional thermocouple. The advantage of the method is that moving parts are not required and hence there is little that can go wrong with the generator. In practice, the container of the radioisotope is the hot junction for the lead telluride or cobalt silicide thermo-electric elements, and the outer surface of the container, which must be thermally insulated from the inner container, serves as the cold junction. One of the most useful nuclides for electrical power generation by this procedure is ^{238}Pu, synthesised from ^{237}Np:

$$^{237}_{93}Np\ (n, \gamma) \qquad ^{238}_{93}Np \quad \xrightarrow[\text{2.1 days}]{\beta} \quad ^{238}_{94}Pu$$

the relatively long half-life (86.4 yr) is particularly attractive, although this means that the heat output is rather low (0.54 W/g). Another useful nuclide is ^{244}Cm (17.6 yr, 2.6 W/g) but ^{242}Cm (163 days, 100 W/g is also of some use for short-term purposes. Otherwise the only uses for the transuranium nuclides are as α-sources, for example, for the production of neutrons by the α-bombardment of an element the isotopes of which have high α, n cross-sections, such as beryllium, and as fission sources, where ^{252}Cf, which has a relatively short spontaneous fission half-life, has found some application.

4. THE DISCOVERY OF ELEMENTS 101–103

The few known isotopes of these elements are all extremely short-lived. Because of this, the only procedure available for their synthesis is cyclotron bombardment of the higher actinides with charged particles. Very sophisticated techniques have to be used for the separation and identification of the products of such bombardments; these have been described in detail elsewhere[6] and the experimental details are not given here.

It is, however, worth mentioning some of the difficulties involved in the identification of the products; the cross-section or probability, σ, for the reactions used to prepare elements 101–103 is extremely small, usually of the order of 0.001 barn (*i.e.* 10^{-27} cm^2). Now the number of atoms, N_A, produced from N_T atoms of a target nucleus when bombarded by an ion beam of intensity I ions per second in a time t, which must be short compared with the radioactive decay half-life of the product nucleus, is given by:

$$N_A = N_T\ I\sigma t$$

To take an example, a microgram of $^{253}_{99}$Es, when bombarded by a helium ion beam of intensity 10^{14} particles/sec/cm^2 would only yield 100 atoms of product, $^{256}_{101}$Md, per second:

$$^{253}_{99}\text{Es} + {}^{4}_{2}\text{He} \rightarrow {}^{256}_{101}\text{Md} + {}^{1}_{0}n$$

In the original work leading to the discovery of mendelevium in 1955, the target amounted to only 10^9 atoms of ^{253}Es and a three-hour irradiation yielded one atom; the final identification of ^{256}Md depended on its elution position (p. 44) in the ion exchange separation of the products. The name of the element is derived from Mendeleev.

The work on element 102 has been rather controversial. The first reported synthesis was carried out in 1957 by a U.S./U.K. team working with scientists at the Nobel Institute for Physics in Sweden; they used carbon (^{13}C) ion bombardment of ^{244}Cm, the suggested reaction being:

$$^{244}_{96}\text{Cm} + {}^{13}_{6}\text{C} \rightarrow {}^{257-x}_{102}\text{No} + x\,{}^{1}_{0}n$$

This result could not be repeated either at the University of California, Berkeley or at Dubna, near Moscow but later work at Berkeley indicated that ^{254}No can be made by ^{12}C bombardment of ^{246}Cm:

$$^{246}_{96}\text{Cm} + {}^{12}_{6}\text{C} \rightarrow {}^{254}_{102}\text{No} + 4\,{}^{1}_{0}n$$

The identification was by the ion-exchange separation of the α-decay product, ^{250}Fm. The Russian work involved oxygen ion bombardment of ^{241}Pu, which seems to produce ^{253}No, and neon ion bombardment of ^{238}U, yielding ^{255}No:

References p. 16

$$^{241}_{94}\text{Pu} + {}^{16}_{8}\text{O} \rightarrow {}^{253}_{102}\text{No} + 4\,{}^{1}_{0}n$$

$$^{238}_{92}\text{U} + {}^{22}_{10}\text{Ne} \rightarrow {}^{255}_{102}\text{No} + 5\,{}^{1}_{0}n$$

Although named nobelium after Nobel by the 1957 workers, later workers can reserve the right to rename the element.

Finally, element 103, lawrencium, named after E.O. Lawrence, the inventor of the cyclotron, was discovered at Berkeley in 1961; ^{257}Lw was produced by bombarding a ^{252}Cf target with boron ions. The reactions involved are:

$$^{252}_{98}\text{Cf} + {}^{10}_{5}\text{B} \rightarrow {}^{257}_{103}\text{Lw} + 5\,{}^{1}_{0}n$$

$$^{252}_{98}\text{Cf} + {}^{11}_{5}\text{B} \rightarrow {}^{257}_{103}\text{Lw} + 6\,{}^{1}_{0}n$$

5. RADIOACTIVITY AND HEALTH HAZARDS

Research on the chemical properties of the shorter-lived actinide nuclides is complicated by the need to guard against accidental ingestion of the nuclides themselves, as well as by the need for shielding the research chemist from the penetrating γ-radiation and neutrons arising from the spontaneous fission of the higher actinides, a mode of decay which becomes important at californium. As an example, ^{252}Cf emits 10^{12} neutrons sec^{-1} g^{-1} as a result of its rather short ($\sim$ 60 yr) spontaneous fission half-life, a level of neutron emission which requires the use of water tanks as neutron shields between the worker and the californium source, and which also requires that all operations on the californium are carried out by remote control.

Most of the more recent research work on actinide chemistry has concerned thorium and uranium, neither of which require shielding or special precautions when the naturally occurring elements (^{232}Th; ^{235}U (0.7%), ^{238}U (99.3%)) are concerned. With protactinium and americium (^{241}Am or ^{243}Am) some degree of local lead shielding may be required to reduce the γ-radiation dose to the worker when the amounts handled exceed about 50 mg; such shielding is not necessary for ^{239}Pu, but may be required for plutonium in which there are appreciable quantities of the heavier isotopes. Neptunium (^{237}Np) presents only an α-hazard, and the problems associated with curium and the elements beyond are usually solved by using micro-techniques, handling only very small quantities of the nuclides concerned.

The main danger results from irreversible α-radiation damage to body tissue following upon ingestion of the radioactive isotope, for the actinides tend to be selectively absorbed in the bone, kidneys and liver, sites where large-scale permanent damage would be fatal. For this reason the maximum permissible body-burden for α-emittters is extremely low, of the order of 0.03 μCi (the amount of the α-emitting nuclide which gives rise to 1100 α-disintegrations per sec). The high toxicity becomes quite apparent if the maximum permissible concentrations per m^3 in air for continuous exposure, assuming a 40-h week, are compared with the corresponding values for carbon monoxide and hydrogen cyanide (Table 2).

TABLE 2

SOME MAXIMUM PERMISSIBLE CONCENTRATIONS IN AIR (mg/m^3)

CO	HCN	^{232}Th	^{238}U	^{237}Np	^{231}Pa	^{242}Cm
100	10	0.27	0.19	4×10^{-6}	4×10^{-8}	3×10^{-11}

From this Table it is clear that the actinides beyond uranium are very much more toxic when ingested than hydrogen cyanide and it is fortunate that these actinides do not form very volatile compounds. The normal method of working with these elements is simply to handle them in glove-boxes, which are merely enclosures in which the floor is used as a laboratory bench. The boxes are maintained under slightly reduced pressure relative to the laboratory, so that, in the event of a leak in the skin of the box, the air flow will be from the laboratory into the box. Boxes for this type of work have been described in numerous reviews and the various types used for differing kinds of radioactive handling have been the subject of a symposium[7].

6. RADIOACTIVITY – CHEMICAL EFFECTS

The α-radiation from any of the actinides will affect the stability of solid compounds, an example being the experimental observation that it is possible to prepare compounds of curium (IV) with ^{244}Cm (half-life 17.6 yr) but not with ^{242}Cm (half-life 163 days), the difference in chemical behaviour being due to the much higher specific α-activity of ^{242}Cm, which leads to faster radiation decomposition than is the case with the longer-lived ^{244}Cm. Similarly, solid plutonium hexafluoride is decomposed by the α-radiation of the plutonium more rapidly than its neptunium analogue, simply because neptunium has a much longer half-life and is therefore less intensely radioactive.

Radiation decomposition of the solvent in solution chemistry only becomes very marked with nuclides of half-lives of less than about twenty years, and then only when milligram or greater quantities of the nuclide are present in volumes of solvent of the order of 1–2 ml. In general, hydrogen peroxide produced by the α-radiolysis of water presents the biggest problem, for where an actinide can exist in several oxidation states, hydrogen peroxide will often reduce the highest and oxidise the lowest.

7. REFERENCES

1   V.M. Goldschmidt, *Geochemistry*, Oxford University Press, London, 1954.
2   P.F. Kerr, *Proc. Intern. Conf. Peaceful Uses of Atomic Energy (1955)*, 9 (1956) 5.
3   C. Frondel, *Proc. Intern. Conf. Peaceful Uses of Atomic Energy (1955)*, 6 (1956) 568.
4   S. Peterson, in G.T. Seaborg, J.J. Katz and W.M. Manning (Eds.), *Natl. Nuclear Energy Series IV-14B*, McGraw-Hill, New York, 1949, pp. 1391, 1393.

5 D. Brown and A.G. Maddock, *Quart. Rev. (London)*, 17 (1963) 289.
6 G.T. Seaborg, *Man-made Transuranium Elements*, Prentice-Hall, New Jersey, 1963.
7 G.N. Walton (Ed.), *Glove Boxes and Shielded Cells*, Butterworth, London, 1958.
8 H.W. Kirby, *Isis*, 62 (1971) 290.

Chapter 2

OXIDATION STATES

1. INTRODUCTION

In the lanthanide series, oxidation states other than +3 are rather unusual and +4 is the highest oxidation state attainable. Only three of the first seven members of the $4f$ series, namely cerium, praseodymium and neodymium, can be oxidised to this higher oxidising agents, cerium(IV) these quadrivalent lanthanides are powerful oxidising agents, cerium (IV) alone being stable in aqueous solution. In contrast to the lanthanide elements, the most stable oxidation states of the actinide elements up to, and including, uranium range progressively from +3 to +6, corresponding to f^o configurations, and for neptunium, plutonium and americium the +6 state is also attained, although oxidation to that state in aqueous solution becomes progressively more difficult as the atomic number increases. There is also evidence for neptunium(VII), plutonium(VII), and, possibly, plutonium(VIII), but at and beyond americium the common oxidation state becomes +3, as in the lanthanides, with trace-scale experiments providing evidence for the existence of the +2 state in aqueous solution for both mendelevium and nobelium. These data are summarised in Table 3.

In compounds of the +2, +3 and +4 oxidation states, the metals are usually present as the simple M^{2+}, M^{3+} or M^{4+} cations or as species ˉ derived from these ions, but for the higher oxidation states the commonest species from uranium onwards, both in solid compounds and in aqueous solution, are the oxygenated cations MO_2^+ and MO_2^{2+} and oxoanions of the type $(MO_6)^{n-}$, where $n = 4, 5$ or 6. A few penta- (p. 105) and hexahalides (p. 120) are also known for uranium, but the only non-oxygenated species recorded for the +5 or higher oxidation states of the elements beyond uranium are the neptunium(V) and plutonium(V) fluorocomplexes (p. 107) and the hexafluorides of these two elements. Pentahalides and related compounds are normal for

TABLE 3

THE KNOWN OXIDATION STATES OF THE ACTINIDE ELEMENTS[a]

Ac	Th	Pa	U	Np	Pu	Am	Cm	Bk	Cf	Es	Fm	Md	No	Lw
					(2)[b]						2	2		
3	(3?)	(3?)	3	3	3	3	3	3	3	3	3	3	3	3
	4	4	4	4	4	4	4	4						
		5	5	5	5	5								
			6	6	6	6								
				7	7									
					8(?)									

[a] The most stable oxidation state in aqueous solution is underlined.
[b] Only known in dilute solid solution in a CaF_2 matrix.

protactinium(V) which does not form the PaO_2^+ ion like the later
actinides. Thus the most stable compounds of the higher oxidation
states from uranium to americium are those formed with difficultly
oxidisable anions, such as F^- or O^{2-}, as is usual with multivalent
elements, and the dioxocations conform to this category. The steps
leading to the formation of the MO_2^+ or MO_2^{2+} ions can be regarded as
consisting of the initial formation of the gaseous M^{5+} or M^{6+} ions, for
which the largest endothermic energy term will be the sum of the first
five or first six ionization potentials, followed by bond formation with
the O^{2-} ion. The product is then either hydrated in aqueous solution or,
in the solid, built into a lattice in which the MO_2^+ or MO_2^{2+} cation
maintains its identity. At the element uranium, the energy contributions
arising from the formation of M-O bonds and the above hydration or
lattice formation are clearly more than enough to balance the ionization
potential requirements. However, between uranium and americium the
thermodynamic stability of the +5 and +6 states, in the shape of the
dioxocations and the hexafluorides, decreases progressively, almost
certainly as a result of increasing ionisation potentials in passing along
the series, and this increase in the endothermic energy term is not
balanced by the increase in hydration or lattice energy as the size of the
actinides decreases.

The moderately low ionisation potential requirements at the beginning of the actinide series and their subsequent increase can be explained in the following way. Along the Periodic Table from the lanthanides to the actinides the $5d$, $6s$ and $6p$ shells are added to the completed $4f$ shell with about a 4–5% increase in radius for any given actinide element as compared with its lanthanide analogue. The slight increase in radius must lead to a slight reduction ($\sim 10\%$) in the effective nuclear charge experienced by the $5f$ electrons, as compared with that experienced by the $4f$ electrons at the corresponding point in the lanthanide series, but the additional screening of the $5f$ electrons from the nucleus by the completed shells mentioned above is probably much more significant in bringing about a reduction in the effective nuclear charge. As a result, the ionisation potentials for the actinides, at the least in the earlier part of the series, should be appreciably lower than those of their lanthanide counterparts. Although ionisation potential data are not available for the actinide elements, so that the proposition cannot be quantitatively demonstrated, the situation is qualitatively comparable with the observed differences in the attainment of high oxidation states in the $3d$, $4d$, and $5d$ transition elements, in which the $4f$ and $5f$ series can be compared in this respect with the $3d$ and $4d$ series. Taking iron, ruthenium and osmium as examples, it can be seen from the ionisation potentials (Table 4) that iron will be very much more difficult to oxidise to Fe(VIII) than ruthenium to Ru(VIII), which, in turn, is more difficult to oxidise to that state than osmium. Likewise, in the d-transition series, the

TABLE 4

IONISATION ENERGIES (eV) OF IRON, RUTHENIUM AND OSMIUM[a]

	I_1	I_2	I_3	I_4	I_5	I_6	I_7	I_8
Fe	7.90	16.18	30.64	57.1	78	102	128	151
Ru	7.5	16.4	28.6	46.5	63	81	100	119
Os	8.7	15	25	40	54	68	83	99

[a]F.A. Cotton and G. Wilkinson, *Advanced Inorganic Chemistry*, Interscience, New York, 2nd Ed., 1966, pp. 797, 910.

decreasing ionisation potentials I_3–I_8 are due to a reduction in the effective nuclear charge experienced by the d electrons resulting from their shielding from the nucleus by the intervening completed electron shells below the d shell which is being filled.

As mentioned earlier, the stability of the higher oxidation states of the actinides decreases in passing along the series from uranium and, at the same time, the ionic radii for any given oxidation state decrease as the atomic number increases. This effect is known as the 'actinide contraction'; the same effect is observed in the lanthanides. Although the decrease in ionic radius for consecutive elements is quite small (Table 5), the cumulative decrease in radius from one end of the series to the other is significantly large. In both the lanthanide and actinide series the contraction in radius results from the successive addition of

TABLE 5

RADII[a] (pm) OF $4f$ AND $5f$ M^{3+} AND M^{4+} IONS

	M^{3+b}	M^{4+}		M^{3+b}	M^{4+}
La	100.6		Ac	107.6	
Ce	98.2	92	Th	–	99
Pr	96.8	90	Pa	–	96
Nd	95.5		U	100.5	93
Pm	94.5		Np	98.6	92
Sm	92.1		Pu	97.4	90
Eu	90.9		Am	96.2	89
Gd	90		Cm	94.6	88
Tb	88.8	84	Bk	93.5	
Dy	87.7		Cf	91.5	
Ho	86.8				
Er	86				
Tm	85.5				
Yb	85.1				
Lu	84.8				

[a]M^{3+} data from J.R. Peterson and B.B. Cunningham, *J. Inorg. Nucl. Chem.*, 30 (1968) 1775; M^{4+} data from G.T. Seaborg, *Man-Made Transuranium Elements*, Prentice-Hall, New Jersey, 1963, p. 93.
[b]Effective octahedral radii.

electrons to the inner, f, shell, for, owing to the shape of the f orbitals (Chapter 12), the shielding of one f electron by another is very poor so that at each increase of atomic number the effective nuclear charge experienced by each f electron increases, and thus the whole f^n shell is reduced in size. Much the same kind of effect is observed in the d transition series; for example, the radii of the Ti^{2+}, Mn^{2+} and Zn^{2+} ions are 0.90, 0.80 and 0.70 Å respectively, the contraction again being due to the increasing effective nuclear charge along the series which results from the rather poor shielding of one d electron by another. Just as we have seen for the actinides, this effect stems from the shapes of the d orbitals, the order of shielding efficiency being $s > p > d > f$.

This rise in the effective nuclear charge will increase the ionisation potentials along the series until, at americium, oxidation states higher than +3 become very difficult to achieve. Thus beyond americium the +3 state is normal and the known compounds of curium(IV) are limited to the dioxide (p. 67) and fluoride (p. 91). Curium is, of course, the analogue of gadolinium, the element which has reached the half-filled f shell stage, and berkelium, which is the analogue of terbium, is more easily oxidised to the +4 state than curium because Bk^{4+} has attained the half-filled shell, f^7, configuration. This half-filled shell effect is, however, more significant in the lanthanides, where it can be used to account satisfactorily for the stabilities of the bi- and quadri-positive oxidation states in that series. The effect derives from the exchange-energy difference between an f^{n+1} and an f^n configuration. This exchange energy is simply the binding energy which arises from the electrostatic interaction exerted on each other by paired electrons with opposed spins.

2. THE STABILITY OF THE +3 OXIDATION STATE

At first sight it may appear somewhat surprising to find that the +3 oxidation state is the most stable oxidation state with respect to oxidation or reduction throughout the lanthanide series and over the latter half of the actinide series, since the outer electron configurations are most commonly $4f^n 6s^2$ and $5f^n 7s^2$ in both series (Table 1, p. 2).

This might well be expected to lead to a stable +2 oxidation state, with the chemistry of the elements consequently following that of the alkaline earths. As the chemistry does not follow this pattern, the +3 state being preferred both in aqueous solution and in the solid state, it is clear that in solution, for example, the oxidation:

$$M^{3+}_{(aq)} + H^{+}_{(aq)} + \tfrac{1}{4} O_{2(g)} \rightarrow M^{4+}_{(aq)} + \tfrac{1}{2} H_2O_{(1)}$$

and the reduction:

$$M^{3+}_{(aq)} + \tfrac{1}{2} H_{2(g)} \rightarrow M^{2+}_{(aq)} + H^{+}_{(aq)}$$

are thermodynamically unfavourable.

The steps in the cycles which lead from the metal to the aquated M^{2+}, M^{3+} and M^{4+} ions are as follows:

$$M_{(s)} \xrightarrow{Lv} M_{(g)} \xrightarrow{I_1 + I_2} M^{2+}_{(g)} + 2\,e^{-} \xrightarrow[\Delta H_{2\,(hyd)}]{} M^{2+}_{(aq)}$$

$$M_{(s)} \xrightarrow{Lv} M_{(g)} \xrightarrow{I_1 + I_2 + I_3} M^{3+}_{(g)} + 3\,e^{-} \xrightarrow[\Delta H_{3\,(hyd)}]{} M^{3+}_{(aq)}$$

$$M_{(s)} \xrightarrow{Lv} M_{(g)} \xrightarrow{I_1 + I_2 + I_3 + I_4} M^{4+}_{(g)} + 4\,e^{-} \xrightarrow[\Delta H_{4\,(hyd)}]{} M^{4+}_{(aq)}$$

where Lv is the latent heat of sublimation of the metal M, I_1–I_4 are the first to fourth ionization potentials, $\Delta H_{2\,(hyd)}$, $\Delta H_{3\,(hyd)}$ and $\Delta H_{4\,(hyd)}$ are the enthalpies of hydration of the M^{2+}, M^{3+} and M^{4+} ions respectively. From these three cycles it is clear that the enthalpies of formation of the aquated ions will be:

$$M^{2+}_{(aq)}, \Delta H_f = Lv + (I_1 + I_2) + \Delta H_{2\,(hyd)}$$

$$M^{3+}_{(aq)}, \Delta H_f = Lv + (I_1 + I_2 + I_3) + \Delta H_{3\,(hyd)}$$

$$M^{4+}_{(aq)}, \Delta H_f = Lv + (I_1 + I_2 + I_3 + I_4) + \Delta H_{4\,(hyd)}$$

so that, in the oxidation of $M^{2+}_{(aq)}$ to $M^{3+}_{(aq)}$, the overall enthalpy of the reaction, ΔH_{ox}, will be equal to the difference between the enthalpies of formation of the two ions:

$$\Delta H_{ox} = I_3 + \Delta H_{3\,(hyd)} - \Delta H_{2\,(hyd)}$$

Consequently the oxidation can be expected to occur if $\Delta H_{3(hyd)} - \Delta H_{2(hyd)} > I_3$. Because the M^{3+} cations are appreciably smaller than the M^{2+}, their heats of hydration will be much greater; taking lanthanum as the nearest example, $\Delta H_{3(hyd)}$ is reported[1] to be -3373 kJ/mole and the La^{2+} ion, if it existed, would have a radius intermediate between that of Sr^{2+} and Ba^{2+}, so that $\Delta H_{2\,(hyd)}$ would be about -1420 kJ. The third ionisation potential of lanthanum[2] is 1850 kJ/g atom, so that ΔH_{ox} for the reaction $La^{2+}_{(aq)} \longrightarrow La^{3+}_{(aq)}$ would be -103 kJ/g ion, quite markedly exothermic. The corresponding free-energy change, ΔG°, is almost equally large and negative, so that the oxidation of $La^{2+}_{(aq)}$ would be spontaneous, as expected. Although accurate hydration enthalpy or free energy data, and experimental ionization potential data, are lacking for nearly all the actinide tripositive ions, the third ionisation potential should increase with increasing atomic number, and this endothermic contribution should be balanced by the increasing hydration enthalpy as the ionic radius M^{3+} decreases across the group. The dipositive state only becomes stable with respect to oxidation at nobelium (No^{2+}, f^{14}) and it must therefore be assumed that I_3 reaches a maximum at this element because of the additional energy required to break into the completed $5f$ shell. Thus, in the lanthanides, I_3 for ytterbium, the homologue of nobelium, is about 385 kJ higher than I_3 for lutetium[3].

In the case of oxidation to the tetrapositive state $(M^{3+}_{(aq)} \longrightarrow M^{4+}_{(aq)})$, the overall enthalpy of the reaction, $\Delta H'_{ox}$, will be:

$$\Delta H'_{ox} = I_4 + \Delta H_{4\,(hyd)} - \Delta H_{3(hyd)}$$

There are no numerical data available for any of these quantities, but on size and charge grounds $\Delta H_{4(hyd)}$ should be about 3200 kJ more exothermic than $\Delta H_{3(hyd)}$ for a given element, which requires I_4 to be numerically about 100 kJ less than this figure for stability with respect to reduction in aqueous solution. This is of the right order of magnitude for cerium in the lanthanides, in which group I_4 is otherwise too large for the $M^{4+}_{(aq)}$ ion to be stable. It is probable that the difficulty of oxidising curium, and the actinides beyond berkelium, to the +4 oxidation state is also entirely due to the high value of I_4.

Similar arguments can be applied to solid compounds. In these it is a combination of the ionisation and lattice energies which leads to the predominance of the terpositive state in the actinides from americium onwards; the requirement, as before, is for the lattice energies of the M^{4+} compounds to exceed those for the M^{3+} compounds by an amount sufficient to balance I_4 in order that the M^{4+} compound may be stable with respect to reduction in the solid state. In the actinides, as mentioned earlier, a few curium(IV) compounds are known, but these are all formed with anions which are not easily oxidised (O^{2-}, F^-).

3. HIGHER OXIDATION STATES IN AQUEOUS SOLUTION

The redox potentials for the various oxidation states of the actinides are reasonably well known and some of the data are given in Table 6. These data show that the couples $M^{4+}_{(aq)}/M^{3+}_{(aq)}$ and $MO^{2+}_{2(aq)}/MO^{+}_{2(aq)}$ are reversible and involve only a single electron change, with neither the breaking nor the formation of metal–oxygen bonds. Such redox changes are kinetically fast when one electron-reducing or -oxidising agents, which do not undergo any bond changes themselves, are used for the oxidation or reduction and then only in the absence of ligands which can complex the cations involved. However, the reaction is not necessarily fast when the oxidation or reduction involves a reagent in which bond making or breaking processes take place. Similarly, the rate of reduction in couples such as $MO^{+}_{2(aq)}/M^{4+}_{(aq)}$ will generally also be slower than in couples where single electron changes are involved.

Considering the $M^{4+}_{(aq)}/M^{3+}_{(aq)}$ systems first, it can be seen that the $U^{3+}_{(aq)}$ ion is very easily oxidised by the aquated hydrogen ion and although this oxidation is normally very fast, it appears that

References p. 32

TABLE 6

FORMAL ELECTRODE POTENTIALS (1 M HClO$_4$) AT 298°K (H$_2$/H$^+_{(aq)}$ = 0V), VOLTS

	$M^{3+}_{(aq)}/$ $M^0_{(aq)}$	$M^{4+}_{(aq)}/$ $M^0_{(aq)}$	$M^{4+}_{(aq)}/$ $M^{3+}_{(aq)}$	$MO^+_{2(aq)}/$ $M^{4+}_{(aq)}$	$MO^{2+}_{2(aq)}/$ $MO^+_{2(aq)}$	$MO^{2+}_{2(aq)}/$ $M^{4+}_{(aq)}$	$MO^{2+}_{2(aq)}/$ $M^{3+}_{(aq)}$
Th		−1.90					
U	−1.80		−0.631	+0.58	+0.063	+0.32	
Np	−1.83		+0.155	+0.739	+1.137	+0.938	+0.677
Pu	−2.03		+0.982	+1.172	+0.913	+1.043	+1.0228
Am	−2.32		(+2.7)	+1.04[a]	+1.60	+1.04	+1.69
Bk			(+1.6)				

[a] $AmO^+_2(aq)/Am^{3+}_{(aq)}$, +1.74 V

Data from G.T. Seaborg, *Man-made Transuranium Elements,* Prentice-Hall, New Jersey, 1963, p. 85.

complexing stabilises uranium(III) sufficiently to allow of the isolation of anionic complex salts from aqueous solution[4]. Neptunium(III) requires only mildly oxidising conditions for conversion to neptunium(IV), atmospheric oxygen being sufficient for the purpose; the potential for the reaction:

$$O_2 + 2 H^+_{(aq)} + 2e^- \longrightarrow H_2O_{2(aq)}$$

is +0.68 V and the $Np^{4+}_{(aq)}/Np^{3+}_{(aq)}$ potential, +0.155 V, is appreciably more negative than this. This means that the reduction of both $U^{4+}_{(aq)}$ and $Np^{4+}_{(aq)}$ to the terpositive state will require moderately strong reducing conditions and a common way of achieving this cleanly is by cathodic reduction. In the case of plutonium, the oxidation of $Pu^{3+}_{(aq)}$ to $Pu^{4+}_{(aq)}$ in aqueous solution is a little more difficult than that of neptunium, hydrogen peroxide in acid solution normally being used, but the reverse reduction to $Pu^{3+}_{(aq)}$ is correspondingly easier, iodide ion or sulphur dioxide being adequate for the purpose. The oxidation of $Am^{3+}_{(aq)}$ to $Am^{4+}_{(aq)}$ in aqueous solution is not, however, possible, for the $M^{4+}_{(aq)}/M^{3+}_{(aq)}$ potential is +2.7 V, which is sufficient to oxidise water; the $M^{4+}_{(aq)}/M^{3+}_{(aq)}$ potential is probably even larger for curium. However, with berkelium, the half-filled shell effect becomes apparent and the oxidation of $Bk^{3+}_{(aq)}$ to $Bk^{4+}_{(aq)}$ is somewhat easier than the oxidation of $Ce^{3+}_{(aq)}$ to $Ce^{4+}_{(aq)}$, so that ceric ion can be used to oxidise $Bk^{3+}_{(aq)}$. The reduction of $Bk^{4+}_{(aq)}$ to the terpositive state is correspondingly more difficult and stannous chloride is commonly used for this purpose.

When the $MO^+_{2(aq)}/M^{4+}_{(aq)}$ potentials are considered, problems arise with both uranium and plutonium, for in this area the aquated ions disproportionate, a process of simultaneous oxidation and reduction:

$$2\,U(V) \longrightarrow U(VI) + U(IV)$$

The overall reaction for uranium(V) is:

$$2\,UO^+_{2(aq)} + 4\,H^+_{(aq)} \longrightarrow U^{4+}_{(aq)} + UO^{2+}_{2(aq)} + 2\,H_2O$$

and it is believed that the following steps are involved:

(i) $UO^+_{2(aq)} + H^+_{(aq)} \longrightarrow (UOOH)^{2+}_{(aq)}$

(ii) $UO^+_{2(aq)} + (UOOH)^{2+}_{(aq)} \longrightarrow UO^{2+}_{2(aq)} + (UOOH)^+_{(aq)}$

(electron transfer)

(iii) $(UOOH)^+_{(aq)} + 3\,H^+_{(aq)} \longrightarrow U^{4+}_{(aq)} + 2\,H_2O$

This reaction sequence can be divided into two parts, an oxidation step, $UO^{2+}_{2(aq)}/UO^+_{2(aq)}$, $E^o = +0.063$ V, and a reduction step, $UO^+_{2(aq)}/U^{4+}_{(aq)}$, $E^o = +0.58$ V, so that, for the complete reaction, $E = (+0.58 - 0.063)$ V $= +0.517$ V. This means that the free-energy change in the reaction, $\Delta G^o\ (= -nFE)$, will be quite large and negative, highly favourable for the reaction.

Putting these observations in general terms, for any disproportionation of the form:

$$2\,M^{x+}_{(aq)} \longrightarrow M^{(x+1)+}_{(aq)} + M^{(x-1)+}_{(aq)}$$

for which the standard potentials E^o ($M^{(x+1)+}_{(aq)}/M^{x+}_{(aq)}$) and
E^o ($M^{x+}_{(aq)}/M^{(x-1)+}_{(aq)}$) are A and B volts respectively, then $\Delta G^o = -F(B-A)$ and if $B > A$, ΔG^o is negative and the reaction is favoured. In other words, if the $M^{x+}_{(aq)}/M^{(x-1)+}_{(aq)}$ potential is more positive than the $M^{(x+1)+}_{(aq)}/M^{x+}_{(aq)}$ potential, disproportionation always results.

Applying this rule to plutonium(V), we find that the $PuO^+_{2(aq)}$ ion is also unstable with respect to disproportionation in aqueous solution, but with plutonium the situation is unusually complex.

It is known that all the four oxidation states between +3 and +6 can co-exist in aqueous solution in equilibrium, and, moreover, that each oxidation state is present at a measurable concentration. From Table 6 it can be seen that plutonium(V) will, like uranium(V) be unstable with respect to disproportionation to plutonium(IV) and plutonium(VI); it is also apparent from Table 6 that the potentials of the systems $Pu^{4+}_{(aq)}/Pu^{3+}_{(aq)}$, $PuO^{2+}_{2(aq)}/Pu^{4+}_{(aq)}$, $PuO^{2+}_{2(aq)}/PuO^+_{2(aq)}$ and $PuO^{2+}_{2(aq)}/Pu^{3+}_{(aq)}$ in $1M$ $HClO_4$ are almost equal. Because of this, it is possible for $Pu^{4+}_{(aq)}$ to disproportionate as well; for example, at low acidity (pH 1 or 0.1 M $HClO_4$), $PuO^+_{2(aq)}$ and $PuO^{2+}_{2(aq)}$ can be formed:

$$3\,Pu^{4+}_{(aq)} + 2\,H_2O \rightleftharpoons PuO^{2+}_{2(aq)} + 2\,Pu^{3+}_{(aq)} + 4\,H^+_{(aq)}$$

$$2\,Pu^{4+}_{(aq)} + 2\,H_2O \rightleftharpoons PuO^+_{2(aq)} + Pu^{3+}_{(aq)} + 4\,H^+_{(aq)}$$

and $\quad 2\,PuO^+_{2(aq)} + 4\,H^+_{(aq)} \rightleftharpoons PuO^{2+}_{2(aq)} + Pu^{4+}_{(aq)} + 2\,H_2O$

Adding the last two equations together leads to the equilibrium:

$$PuO^+_{2(aq)} + Pu^{4+}_{(aq)} \rightleftharpoons PuO^{2+}_{2(aq)} + Pu^{3+}_{(aq)}$$

In 0.1 M $HClO_4$ and unit ionic strength at $25°$, the equilibrium constant

$$K = [PuO^{2+}_{2(aq)}]\ [Pu^{3+}_{(aq)}]/[PuO^+_{2(aq)}]\ [Pu^{4+}_{(aq)}]$$

has been found[5] to be 10.7. Consequently, irrespective of its initial oxidation state in aqueous solution, provided that this is greater than +3, plutonium will appear in all four oxidation states in weakly acid solutions when complexing ligands are absent.

Although $Pu^{4+}_{(aq)}$ is just stable with respect to disproportionation into $PuO^{2+}_{2(aq)}$ and $Pu^{3+}_{(aq)}$ on the foregoing criteria, the molal entropy of $Pu^{4+}_{(aq)}$ is large and negative because of the high charge and relatively small size of the ion. There is therefore a large ΔS^o contribution in the disproportionation reaction of $Pu^{4+}_{(aq)}$, of the order of 430 J/°/mole, so that a temperature increase of about 45° will increase the equilibrium constant by a factor of about 1000. The disproportionation will also be favoured by bases, provided that the pH is not increased to the point at which the highly insoluble plutonium(IV) hydroxide precipitates[6].

Phillips and Williams[7] handle electrochemical data of this kind very neatly in the form of a plot of "volt-equivalents" against oxidation state. The "volt-equivalent", $-nE^o$, is equivalent to $\Delta G^o/F$, and the actinide data from Table 6 are plotted in this form in Fig. 3. From the uranium curve in this figure, it is apparent that when the curvature is convex upwards disproportionation occurs, a simple way of detecting this reaction possibility. From Fig. 3 it is also clear that $PuO^+_{2(aq)}$ and $Am^{4+}_{(aq)}$ will also disproportionate, and that $U^{4+}_{(aq)}$, $Np^{4+}_{(aq)}$, $NpO^+_{2(aq)}$ and $AmO^+_{2(aq)}$ will be stable with respect to disproportionation in 1 M HClO$_4$, a result which is equally self-evident on the criteria discussed earlier (p. 28). The $Am^{4+}_{(aq)}/Am^{3+}_{(aq)}$ potential is so large that it can also oxidise water, providing an alternative reaction route.

Although the electrode potential for the Pa(V)/Pa(IV) system is not known accurately[8], it seems to be more negative than that for the $Np^{4+}_{(aq)}/Np^{3+}_{(aq)}$ couple, so that oxidation of protactinium(IV) in aqueous solution by oxygen is quite rapid. A selection of the oxidising agents which can be used for the production of the higher oxidation states of the other actinides (up to MO^{2+}_2) in aqueous solution is given in Fig. 4.

Neptunium(VI) and plutonium(VI) are oxidised in alkaline media to the septavalent state by ozone and by persulphate, hypochlorite and perxenate ions. Solutions of the septavalent elements are deep green and salts of the $[MO_5]^{3-}$ (M = Np, Pu) ions that are present in the alkaline solution can be isolated quite easily. For example, the green compound $Ba_3(NpO_5)_2 \cdot n\, H_2O$ and the brownish black $Ba_3(PuO_5)_2 \cdot n\, H_2O$ are

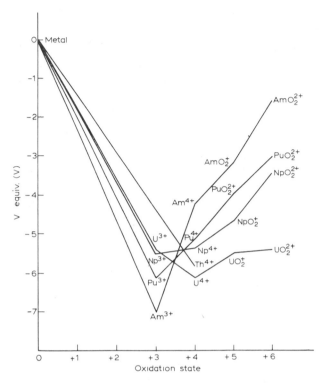

Fig. 3. Oxidation/reduction data for the actinides; a plot of volt-equivalents
$(-nE^\circ = \Delta\ G/F)$ for the actinides, after Phillips and Williams[7] (Fig. 22.4, p. 134),
by courtesy of Clarendon Press, Oxford.

precipitated on the addition of a solution of a barium salt
to the alkaline solution of the septavalent actinide element[8]. These
anions appear to be derived from the acidic species $MO_2(OH)_3$ in
which the actinyl, MO_2, group is present. Plutonium(VII) is an
appreciably more powerful oxidising agent than neptunium(VII), as
would be expected. The highest concentration of neptunium(VII) in
alkaline solution recorded to date is about 12 g/l, obtained by ozone
oxidation of neptunium(VI) in 2 M lithium hydroxide; the maximum
concentration obtained for 2 M sodium hydroxide solution under similar
conditions is only[9] 5 g/l.

The reported[10] formation of americium(VII) by ozone oxidation
of a suspension of americium(V) hydroxide in 3 N sodium hydroxide is

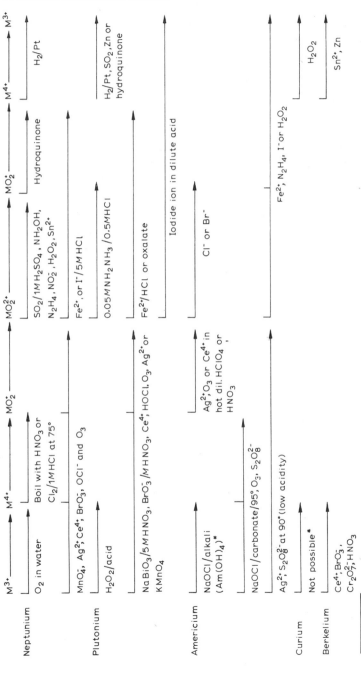

* The only aqueous solutions that can be prepared are obtained by dissolving $Am(OH)_4$ or CmF_4 in 15 M NH_4F or CsF respectively

Fig. 4. Oxidation and reduction of the actinides in aqueous media (K. W. Bagnall, *Sci. Progr.*, 52 (1964) 76). Reproduced by permission of the publishers, Blackwell, Oxford.

erroneous[11]; the violet colour, originally ascribed to americium(VII), is apparently due to the formation of iron(VI).

The solid compounds isolated from aqueous solution, and those prepared by dry methods, are discussed later (p. 73).

4. REFERENCES

1 K.B. Harvey and G.B. Porter, *Introduction to Physical Inorganic Chemistry*, Addison-Wesley, Reading/Palo Alto/London, 1965, p. 326.
2 J. Sugar and V. Kaufman, *J. Opt. Soc. Am.*, 55 (1965) 1283.
3 M.M. Faktor and R. Hanks, *J. Inorg. Nucl. Chem.*, 31 (1969) 1649.
4 R. Barnard, J.I. Bullock and L.F. Larkworthy, *Chem. Commun.*, (1968) 960.
5 M. Kasha, in G.T. Seaborg, J.J. Katz and W.M. Manning (Eds.), *The Transuranium Elements*, Natl. Nucl. Energy Series, Vol. IV–14B, Mc.Graw-Hill, New York, 1949, p. 295.
6 D.A. Johnson, *Some Thermodynamic Aspects of Inorganic Chemistry*, Cambridge University Press, London, 1968, p. 85.
7 C.S.G. Phillips and R.J.P. Williams, *Inorganic Chemistry*, Vol. 2, Clarendon, Oxford, 1965, p. 134.
8 V.I. Spitsyn, A.D. Gel'man, N.N. Krot, M.P. Mefodiyeva, F.A. Zakharova, Yu.A. Komkov, V.P. Shilov and I.V. Smirnova, *J. Inorg. Nucl. Chem.*, 31 (1969) 2733.
9 V.I. Blokhin, N.N. Krot and A.D. Gel'man, *Soviet Radiochem.*, 12 (1970) 764.
10 V.P. Zaitseva, *Dokl. Chem.*, 188 (1969) 802.
11 V.P. Zaitseva, *Radiokhimiya*, 13 (1971) 658.

Chapter 3

SEPARATION AND PURIFICATION

1. INTRODUCTION

With the development of an increased demand for thorium and uranium for nuclear energy, the processes which have been devised for their production on an industrial scale, and for the extraction of the transuranium elements from irradiated uranium, have increased in number and complexity to such an extent that only a very superficial review can be attempted in the space available. A more detailed treatment of the processes used for the separation of individual actinide elements up to 1957 has been given by Katz and Seaborg[1] and later developments appear in the annual reviews on the progress in nuclear technology[2].

Separation and purification processes can be conveniently divided into four groups; the first comprises the separation of natural thorium and uranium from their ores and the second, the isolation of actinium and protactinium from uranium ore residues. The third group, in which differences in the stabilities of the higher oxidation states are exploited, covers the separation of neptunium and plutonium from neutron-irradiated uranium, while the fourth group covers the isolation of the actinides from americium to lawrencium.

2. THORIUM

The first step in the production of thorium from monazite (p. 4) is the treatment of the mineral, as inexpensively as possible, in order to obtain a suitable solution. This is achieved either by digestion with hot concentrated sulphuric acid, an excess of which is necessary for the complete dissolution of the relatively insoluble thorium phosphate present; or by treatment with hot concentrated sodium hydroxide solution, which converts the phosphates to the insoluble hydroxide. The problem then consists in devising a separation of thorium from the lanthanides and uranium.

References p. 46

A partial separation of thorium and uranium(VI) from the lanthanide elements is relatively easy, provided that any cerium present is
in the terpositive state; in order to ensure this, hydrogen peroxide or
sodium nitrite is used to reduce the cerium(IV). The addition of an
excess of ammonium oxalate or carbonate to the diluted sulphuric acid
leach-solution will precipitate a large part of the lanthanides as the
insoluble oxalates or carbonates, both thorium and uranium(VI)
forming water-soluble oxalato- or carbonato-complexes provided a large
enough excess of these reagents is present. However, as their atomic
number increases, that is, as the ionic radius decreases, the lanthanides
become increasingly more easily complexed by oxalate and carbonate so
that the method is suitable only as a primary concentration step. An
alternative to this complexing procedure is to add aqueous ammonia to
the sulphuric acid leach-solution, adjusting to pH 1; this precipitates
thorium, uranium(VI) and 3–5% of the lanthanides. A better separation
from the lanthanides can be achieved if the hydrolysis of hexamethylene
tetramine is used as the source of ammonia for the precipitation; this
reagent is generally used for the separation of thorium for analytical
purposes and not industrially.

When a caustic-leach process is adopted, the leach liquor is
adjusted to pH 5.8, which leaves a mixture of insoluble thorium,
uranium and lanthanide hydroxides as before. From this point
the best procedure for the final separation is to dissolve the residue in
nitric acid and to extract thorium and uranium(VI) into a solution of
tributyl phosphate (TBP) in kerosene or into methylisobutyl ketone
(hexone), leaving the lanthanides (provided that cerium is present as
cerium(III)) in the aqueous phase. However, the conditions of the TBP
extraction step can be adjusted so that thorium and uranium are
separated from each other, for example, by varying the rates of flow of
the organic and aqueous phases in a multiple stage solvent-extraction
unit. Alternatively, if the separation of thorium and uranium is not
effected at this stage, the two elements are recovered from the organic
phase by washing it with very dilute acid or with an ammonium
carbonate solution; the wash liquor is adjusted to 4–8 M in nitric acid
and 0.05 M in ammonium fluoride. After this treatment a further TBP

extraction cycle then removes the uranium(VI), leaving the thorium in the aqueous phase.

A wide variety of other solvent extraction systems have been investigated with varying success; the best of these involve chelating agents such as 8-hydroxyquinoline (oxine), nitrosophenylhydroxylamine (cupferron), thenoyltrifluoroacetone (TTA) and acetylacetone; details of these systems have been given in earlier books[1, 3].

Chromatographic separation of thorium from uranium and the lanthanides on alumina and cellulose columns is also practicable[4], but ion-exchange procedures do not seem to have been applied industrially for the separation of thorium. Thorium(IV) is, as one might expect for a highly charged cation, strongly sorbed on cation resins from acid solution, and it is not eluted to any appreciable extent from the resin by washing with nitric or hydrochloric acid, so that all other cations can be removed quite easily. This leaves the thorium to be eluted by complexing anions such as oxalate.

3. URANIUM

Many uranium ores are highly complex mixtures, which may contain 30–40 other elements and, in some instances, organic matter; a detailed description of the various chemical treatments which have been devised for the different ores would be of little interest here and only the general principles involved in these operations are discussed.

An initial concentration step is necessary for ores of low uranium content in order to avoid wasting reagents at the leaching stage. This can be done by making use of the high density of the uranium-containing part of the ore, the concentration being effected by suspending the crushed ore in water and collecting the denser sediment. The concentrates, or the rarer, richer ores which do not require this treatment, are roasted at about 800° to facilitate subsequent treatment and to eliminate organic material as far as possible. The roasted material is then leached, either with sulphuric acid or, if the accompanying non-uraniferous minerals are attacked and would thus consume an excessive amount of the acid, with aqueous sodium carbonate, in which the very stable uranyl tricarbonato-anion $[UO_2(CO_3)_3]^{4-}$, is formed. In both cases an oxidising agent must be added if any uranium(IV) is present.

The aqueous solution resulting from either treatment is usually filtered to remove fine particulate matter, after which the uranium is separated by one of the established procedures: chemical precipitation. solvent extraction or anion exchange. Chemical precipitation is preferred for the recovery of uranium from the carbonate leach solution; the addition of sodium hydroxide to this solution raises the pH to the point at which uranium is precipitated as a uranate or polyuranate. The unavoidable excess of sodium hydroxide is subsequently treated with carbon dioxide and the sodium carbonate solution is passed back to the leaching stage. Alternatively, hydroxyl ion can be generated in the leach solution electrolytically to bring about the precipitation. Another precipitation method takes advantage of the insolubility of uranium(IV) in carbonate solution; for this purpose the uranium(VI) in the solution is reduced to uranium(IV) by hydrogen under pressure in the presence of a recoverable nickel catalyst.

Anion exchange techniques are used for the separation of uranium from both acid and alkaline leach liquors, the uranium being eluted from the resin with nitric acid. As described for thorium, solvent extraction can be used to isolate uranium from nitric acid solution. However, a more convenient procedure for the sulphuric acid leach-solution is to add a thiocyanate and extract the resulting uranyl thiocyanate into TBP, recovering the uranium from the organic phase by backwashing it with aqueous carbonate.

4. ACTINIUM

Most of the actinium present in uranium ores remains with the lanthanide fraction during the extraction of the uranium; this is because the radii of the terpositive actinium, lanthanum and earlier lanthanide ions are rather similar and there is little difference in their chemical behaviour. This makes the separation of actinium from uranium ore residues rather difficult; a classical separation method was multiple fractional crystallisation of the mixed actinium/lanthanide double nitrates of magnesium or manganese, species such as $2 M (NO_3)_3 \cdot 3 Mn (NO_3)_2 \cdot 24 H_2O$ being involved, as in the older procedure used for the separation of the lighter (cerium group, $Z = 57\text{-}62$) lanthanides. Another obsolete method involved the fractional precipitation of the hydroxides with carbonate-free ammonia, actinium hydroxide being somewhat more soluble than lanthanum hydroxide. These now largely discarded methods have been discussed in more detail elsewhere[5]. Nowadays the separation and purification of actinium from uranium ores or ore residues is more conveniently achieved by ion exchange, the method in general used for the separation of the lanthanides and of the later, predominantly terpositive, actinides. This procedure is discussed in more detail later (p. 42).

As mentioned earlier (p. 6), it is possible to synthesise ^{227}Ac from ^{226}Ra; the primary separation of actinides from the neutron-irradiated radium is achieved by extraction from aqueous solution (pH 6) into thenoyltrifluoroacetone (TTA); from this the actinium is recovered by washing the organic phase with $6 N$ HCl. The actinium is then purified either by further solvent-extraction cycles with TTA, under conditions of carefully controlled pH, or by ion exchange. In the final stage of each, the actinium is precipitated as oxalate.

5. PROTACTINIUM

One of the main problems encountered in isolating protactinium from uranium-ore residues is the difficulty experienced in keeping the element in aqueous solution. In the absence of complexing anions such as fluoride or sulphate, protactinium(V) undergoes hydrolysis with the formation of insoluble, polymeric species. Indeed, only one large-scale isolation of protactinium has been recorded; in this about 130 g of the element was recovered from a uranium extraction residue known as "ethereal sludge". This sludge separated at the stage when uranium was extracted from a nitric acid leach solution by ether, as practised in one of the older uranium recovery processes. It contained about 3 g of protactinium per ton, representing about a ten-fold enrichment in protactinium as compared with the equilibrium concentration in natural

uranium (p. 6). The "sludge" was treated with 4 M nitric acid–
0.5 M hydrofluoric acid to leach out uranium and protactinium and
the uranium was extracted into TBP. The addition of aluminium chloride
to the remaining aqueous phase removed fluoride ion, and in the absence
of any complexing ion the protactinium is precipitated along with a mixture of hydrolysed species and phosphates. The phosphate ion was eliminated by treatment with sodium hydroxide, and the resulting hydroxides
or hydrous oxides were separated and dissolved in a mixture of
hydrochloric and hydrofluoric acids. The protactinium was recovered
from this solution by a series of solvent-extraction steps in which
diisobutyl ketone was used. The final stage consisted of anion exchange
from hydrochloric acid solution. The product thus obtained was
contaminated with about 4% of niobium, from which the protactinium
can be separated by a precipitation of the peroxide. Other available
separation processes have been reviewed elsewhere[6].

6. NEPTUNIUM AND PLUTONIUM

The isolation of neptunium and plutonium from neutron-irradiated
uranium involves the separation of these three elements from each other
and from the fission products, which include the lanthanides. All of the
processes, which have been described in more detail in the literature[1],
depend on the different stabilities of the higher oxidation states of
uranium, neptunium and plutonium in aqueous solution. When neutron-
irradiated uranium is dissolved in nitric acid, uranium is oxidised to
UO_2^{2+}, neptunium is normally present as NpO_2^+ and plutonium in the
+4 state, probably as the $Pu(NO_3)_6^{2-}$ anion when the concentration of
nitrate ion is sufficiently high. Nitrite is often added to the solution in
the dissolver in order to reduce any neptunium(VI) to the +5 state and
any plutonium(VI) to the +4 state. The separation can then be achieved
by solvent extraction or by a chemical precipitation process.

Described in outline, one solvent-extraction procedure consists in
extracting uranium(VI) and plutonium(IV) together into TBP as the
complexes $UO_2(NO_3)_2 \cdot 2$ TBP and $Pu(NO_3)_4 \cdot 2$ TPB, leaving
neptunium(V) in the aqueous phase along with the fission products.

The separated organic phase is washed with an aqueous solution of a mild reducing agent, such as potassium iodide, whereupon the plutonium is reduced to the +3 state and passes from the organic phase into the aqueous solution. The uranium is recovered from the organic phase as described earlier (p. 34). The original aqueous phase, containing neptunium(V) and fission products, is then treated with an acidified bromate solution and heated to 95°; by this means neptunium(V) is oxidised to NpO_2^{2+}, which is then extracted from the aqueous phase with TBP. It is recovered from the organic phase by washing with a reducing agent, such as sodium nitrite, which converts neptunium(VI) back to the +5 state. This separation procedure is outlined in flow-sheet form in Fig. 5.

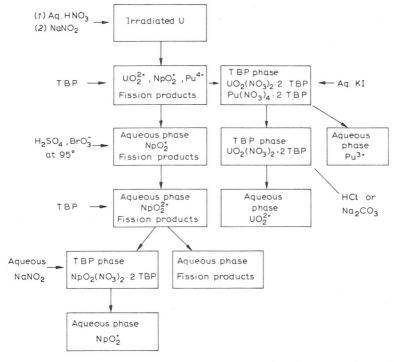

Fig. 5. A solvent extraction scheme for the separation of uranium, neptunium and plutonium from neutron irradiated uranium.

References p. 46

An alternative to the simplified solvent extraction scheme is to make use of the insolubility of the actinide tri- and tetra-fluorides. The mixture of uranium(VI), neptunium(V) and plutonium(IV) is treated with sulphur dioxide, which does not affect uranium(VI) but reduces the last two to the +4 and +3 states respectively. Addition of fluoride ion then precipitates NpF_4, PuF_3, the fission-product lanthanide trifluorides and fission-product zirconium tetrafluoride, leaving the uranium(VI) in solution from which it can be recovered by solvent extraction with TBP or ether.

The fluorides in the mixture are converted to hydroxides with aqueous alkali, dissolved in sulphuric acid, and treated with bromate at $95°$ in order to oxidise neptunium and plutonium to the +6 state. Addition of fluoride ion to this solution precipitates all the remaining fission products, leaving neptunium(VI) and plutonium(VI) in the filtrate; treatment of the filtrate with sulphur dioxide leads to the precipitation of NpF_4 and PuF_3. After reconversion of the fluorides to hydroxides and their redissolution in sulphuric acid, treatment with aqueous bromate at room remperature causes oxidation of the elements to neptunium(VI) and plutonium(IV) respectively. Addition of fluoride now precipitates PuF_4, leaving neptunium(VI) in solution, so that the separation has been achieved. Fig. 6 shows the complete separation in outline. Chemical separation methods of this kind are time-consuming and are not commonly used on a large scale, for which solvent-extraction procedures are preferred.

The last stages in the purification of these elements are often by ion-exchange methods; thus with neptunium, in which the main impurity is usually plutonium, anion exchange is employed. One suitable system consists in reducing neptunium to the +4 state and plutonium to the +3 state, usually with iodide, in moderately concentrated hydrochloric acid. This solution is passed down an anion exchange column, on which the neptunium, present as $NpCl_6^{2-}$, is sorbed, while the plutonium passes on through the column. The neptunium can be eluted from the column with a dilute acid wash.

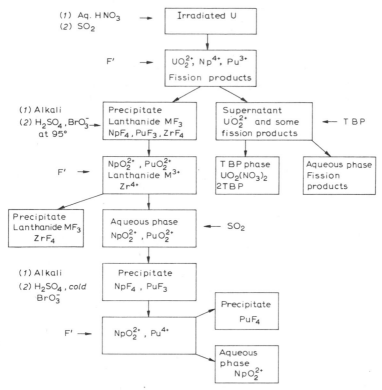

Fig. 6. A possible chemical procedure for the separation of uranium, neptunium and plutonium from neutron irradiated uranium.

7. AMERICIUM AND HIGHER ACTINIDES

From americium to the end of the actinide series the normal oxidation state of the elements in aqueous solution is +3; the single exception is nobelium, for which the +2 state appears to be very stable with respect to oxidation[7]. Of these elements, only americium, berkelium and, possibly, californium can be oxidised beyond the +3 state in aqueous solution; this property of americium and berkelium can be utilised in the purification of these two elements. In the case of americium, oxidation of the terpositive element with ozone or perdisulphate in carbonate solution precipitates the americium(V) carbonato-complex, leaving the lanthanides and higher actinides in solution, together with any uranium, neptunium or plutonium. Similarly, oxidation of berkelium to the +4 state in nitric acid could

provide a basis for the solvent-extraction separation of this element
from other terpositive actinides.

It is also worth mentioning that a convenient source of americium
(^{241}Am) is the plutonium parent, ^{241}Pu. This may be present in
substantial quantities in plutonium which has been subjected to a
prolonged neutron irradiation, or in plutonium separated from long
irradiated uranium, which amounts to the same thing. Because ^{241}Pu
has a half-life of only 14.03 yr[8], it is possible to store it until
sufficient ^{241}Am has grown into the material for separation to be
worthwhile. Americium can be readily separated from plutonium, since
in 8 N nitric acid the plutonium is in the +4 state as the $Pu(NO_3)_6^{2-}$ ion
and the americium remains in a cationic form. When such a solution
comes into contact with an anion exchange resin, the plutonium is
sorbed virtually quantitatively and americium passes through the ion
exchanger. An extension of this procedure consists in leaving the
plutonium on the resin until more americium has grown in from the
decay of ^{241}Pu. The americium can then be washed off the resin with
8 N nitric acid as before; this arrangement has been termed "an
americium cow", the removal of the americium corresponding to a
"milking". Similarly[14], ^{249}Cf can be milked from its parent, ^{249}Bk.

Individual members of the transplutonium actinides or of the
lanthanide series can be separated from other members of the same
series by ion-exchange techniques. From Table 5 (p. 21) it can be seen
that the terpositive ionic radii in each series decrease quite regularly as
the atomic number increases; it is the slight change in chemical
properties which follows from a decrease in radius that provides the
basis for separations of this kind, for the size and charge of a cation
determine its attraction, or lack of attraction, for electrons or anions,
that is to say, they determine the basicity of the cation. Thus large
cations of low charge have very little attraction for electrons and are
basic in character (*e.g.* Cs^+, 1.69 Å), whereas small cations of high
charge have a considerable attraction for electrons and therefore do not
exist normally as cationic species (*e.g.* Al^{3+}, 0.50 Å). The terpositive
actinide and lanthanide ions are appreciably larger than the majority of
the d-transition element terpositive ions, so that the actinide or

lanthanide M^{3+} ions are much more ionic in character than the d transition element M^{3+} ions. As the ionic radius decreases with increasing atomic number, the basicity of the ion also decreases and it is this which provides the key to the separation, for the smaller the cations in either f transition series, the more readily are they complexed.

As a general rule the ion-exchange separation consists in adding a solution of the higher actinides (or any of the lanthanides) in dilute acid to a cation-exchange resin, whereupon all the M^{3+} ions are sorbed on the resin. The separation occurs at the elution stage, when the ions are washed off the resin with a solution of a complexing anion, such as α-hydroxybutyrate, citrate or ethylenediamine tetraacetate (EDTA). Then the ions of smallest crystal radius, that is, those which are the least basic and hence the most easily complexed ions of the series, wash off the resin first, followed in order by the next smallest cation and so on. Thus the elution order is the reverse of the order of the atomic numbers and follows the order of increasing crystal ionic radius. Since we are dealing with aqueous solutions, we can also relate the elution order to the radii of the hydrated ions; because the tendency for ions of a given charge to hydrate increases as the crystal radius decreases, the ions of smallest crystal radius will tend to have the largest hydrated ionic radius. The elution order can then be described as being in order of decreasing hydrated ionic radius[9]. Typical actinide and lanthanide elution orders are shown in Fig. 7.

As explained above, the terpositive ionic radii govern the elution so that in a mixture of lanthanide and actinide elements there will be some overlap in the elution of elements of similar radii in the two series; for example, Am^{3+} overlaps with Pm^{3+}, and Cm^{3+} with Sm^{3+}. Consequently the method described above can be used only to separate the actinides from one another when lanthanides are absent; it will not necessarily separate the actinides from the fission-product lanthanides. This problem is overcome by making use of the fact that the terpositive actinides are much more readily complexed than the lanthanides in concentrated (13.3 M) hydrochloric acid or in ethanol saturated with hydrogen chloride[10, 11].

This difference between the complexing behaviour of the

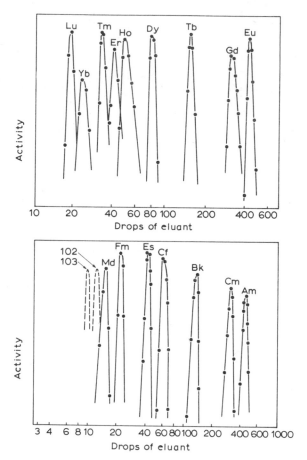

Fig. 7. The elution of terpositive lanthanide and actinide ions from Dowex-50 ion-exchange resin with ammonium α-hydroxyisobutyrate. The predicted positions for elements 102 and 103 are indicated by dotted lines (Glenn T. Seaborg, *Man-Made Transuranium Elements*, (C) 1963, p. 49). Reprinted by permission of Prentice-Hall, Englewood Cliffs, N.J. (U.S.A.).

terpositive lanthanides and actinides arises from the 5*f* orbitals being more accessible for bonding than are the 4*f* orbitals in the lanthanides. The 5*f* orbitals extend spatially into the 6*d* and 7*s* orbital regions and are much more poorly shielded from external fields by the outerlying 7*s* electrons than are the 4*f* orbitals, which do not extend appreciably into the $5d$ and $6s$ regions after cerium and are well below these regions for elements later in the series.

In the actinide/lanthanide separation, the mixture of terpositive ions is sorbed on a cation-exchange resin as before; they are then desorbed with concentrated hydrochloric acid, whereupon the more easily complexed actinides are eluted as a group well ahead of the lanthanides. This procedure does not separate the actinides cleanly from each other, but the actinide fraction obtained in this way can be treated subsequently by the ion-exchange procedure described earlier (p. 43).

8. ANALYTICAL CHEMISTRY

The analytical chemistry of thorium[3], uranium[12] and of the actinides[13] in general has been the subject of recent books; this aspect of the chemistry of the actinides is therefore treated very briefly here.

Thorium, protactinium and uranium are commonly determined by weighing the oxides, ThO_2, Pa_2O_5 and U_3O_8 respectively, obtained by ignition of the hydroxides, oxalates or other suitable compounds in air. Care must be taken in the case of uranium because too high a temperature of ignition ($> 1050°$) leads to the formation of the dioxide, and ignition below $800°$ leaves some uranium trioxide. A number of volumetric methods[3, 12] are also available for the determination of thorium and uranium, the former being commonly estimated by means of EDTA or its derivatives; the latter is more easily estimated by oxidising uranium(IV) to uranium(VI) with dichromate.

Because all of the known actinide nuclides are radioactive, and most of those used in chemical studies emit α-particles, the normal method of detecting and determining them is by counting the emitted α-particles. This requires a knowledge of the half-life of the nuclide involved, and if the element is not mono-isotopic, then the isotopic composition and half-lives of each isotope of the element present must be known. For example, neptunium is available as isotopically pure [237]Np and presents no problem, whereas plutonium may be a mixture of isotopes ranging in mass from 238 to 242, the isotopic composition depending upon the total neutron irradiation. The weight of the element present can then be calculated from the radioactive emission; further details are available elsewhere[13].

References p. 46

The α-spectrum of a sample of the actinide which is to be estimated is also useful as an additional check on its purity; this is because most α-emitting nuclides emit α-particles of different energies so that the spectrum obtained by means of a high resolution α-spectrometer can provide useful information about the presence or otherwise of other α-emitting impurities.

However, actinium (^{227}Ac) is a more difficult element to determine radiometrically because of the very low energy of its emitted β-particle; this makes a specimen of pure actinium appear almost non-radioactive immediately after its purification, but the decay products (^{227}Th and the nuclides in the decay chain which follows) are more easily detected and hence most methods of determining actinium are based on counting ^{227}Th, ^{223}Fr or ^{211}Pb, all of which occur in the ^{227}Ac decay chain[5].

9. REFERENCES

1 J.J. Katz and G.T. Seaborg, *The Chemistry of the Actinide Elements*, Methuen, London, 1957.
2 *Progress in Nuclear Energy, Process Chemistry*, Reviews published annually, Pergamon, Oxford.
3 D.I. Ryabchikov and E.K. Gol'braikh, *The Analytical Chemistry of Thorium*, Pergamon Press, Oxford, 1963.
4 A.F. Williams, *Analyst*, 77 (1952) 297.
5 K.W. Bagnall, *The Chemistry of the Rare Radioelements*, Butterworth, London, 1957.
6 D. Brown and A.G. Maddock, *Quart. Rev. (London)*, 17 (1963) 289.
7 K. Hulet, in U.S.A.E.C. unclassified document UCRL-18667.
8 F.L. Oetting and S.R. Gunn, *J. Inorg. Nucl. Chem.*, 29 (1967) 2659.
9 G.E. Boyd, J. Schubert and S.W. Adamson, *J. Am. Chem. Soc.*, 69 (1947) 2818.
10 K. Street and G.T. Seaborg, *J. Am. Chem. Soc.*, 75 (1950) 2790.
11 R.M. Diamond, K. Street and G.T. Seaborg, *J. Am. Chem. Soc.*, 76 (1954) 1461.
12 D.I. Ryabchikov and M.M. Senyavin (Eds.), *The Analytical Chemistry of Uranium*, Israel Program for Scientific Translations, Jerusalem, 1963.
13 A.J. Moses, *The Analytical Chemistry of the Actinide Elements*, Pergamon, Oxford, 1963.
14 B.F. Myasoedov and K.V. Barsukova, *Radiochem. Radioanalyt. Letters*, 6 (1971) 275.

Chapter 4

THE METALS

1. INTRODUCTION

All of the actinide elements are metals, as would be expected from their electron configurations (Table 1, p. 2) and quite large atomic radii (Table 7).

TABLE 7

THE ATOMIC RADII OF SOME $4f^a$ AND $5f^b$ ELEMENTS

Element	Radius, pm	Element	Radius, pm
La	187.7	Ac	188
Ce	182	Th	179
Pr	182.8	Pa	163
Nd	182.1	U	156
Pm	(?)	Np	155
Sm	180.2	Pu	159
Eu	204.2	Am	173
Gd	180.2	Cm	174
Tb	178.2	Bk	170

[a] T. Moeller, *The Chemistry of the Lanthanides*, Reinhold, New York, 1963, p. 20.
[b] From X-ray crystallographic data (Table 8).

It is immediately obvious from Table 7 that the atomic radii of the first six members of the lanthanide series decrease almost regularly from lanthanum to gadolinium with the exception of europium, the atomic radius of which is substantially greater than that of the other members. This is because europium has only two electrons available for metallic bonding as against three electrons for the elements from lanthanum to samarium, a consequence of the special stabillity of the half-filled f shell (f^7). Ytterbium, the penultimate member of the lanthanide series, also shows this increased atomic radius, and for a similar reason, in this case because of the additional stability of the completed f shell (f^{14}). As a result, these two lanthanide metals closely resemble the alkaline earth metals in both physical and chemical properties, including solubility in liquid ammonia, in which they form dark-blue solutions.

It is equally obvious from Table 7 that the atomic radii of the actinide metals do not show any similar trends and the very sharp decrease in metallic radius between actinium and uranium is similar to that observed in the d transition metals, suggesting that there are no $5f$ electrons present in these earlier actinide metals; it is quite probable that no more than one $5f$ electron is present even in neptunium metal[1]. It is also notable that the atomic radii are appreciably smaller than those of their lanthanide counterparts, whereas the terpositive ionic radii are somewhat larger than those of the lanthanides. This difference between elements of the two f series appears to be due to the availability of more than three electrons for metallic bonding in each of the earlier actinide metals. Since the atomic, or metallic, radius decreases from actinium to uranium, and then remains fairly constant to plutonium, it is reasonable to assume that thorium, protactinium and uranium, for which the most stable oxidation states increase from +4 to +6, contribute increasing numbers of electrons to the metallic bond in that order. Neptunium and plutonium, with atomic radii similar to that of uranium, contribute much the same number of electrons as uranium, whereas americium, although capable of oxidation to the +6 state, seems only to contribute close to 3 electrons and so can reasonably be regarded as the first lanthanide-type member of the actinide series. Some suggested values for the electron contributions, expressed as "metallic valencies", have been given by Cunningham and Wallmann[2]. It is, however, interesting to note that the atomic radius of americium differs little from that of curium, in contrast to the observed difference between the atomic radii of europium and gadolinium, so that it is hardly surprising to find that americium metal does not dissolve in liquid ammonia and does not form stable compounds containing the Am^{2+} ion.

The actinide reduction potentials, $M^{3+}_{(aq)}/M^0_{(s)}$, in the early actinides also differ in magnitude and general trend from those of the corresponding lanthanides. Although the actinides, like the lanthanides, are highly electropositive, the $M^{3+}_{(aq)}/M^0_{(s)}$ potentials decrease from -1.80 V for uranium to -2.32 V for americium (Table 6, p. 26) which is the opposite of the behaviour in the lanthanide series in which this potential changes remarkably smoothly from -2.52 V at lanthanum to -2.27 V at lutetium. This difference must be related to the non-lanthanide trend discussed above and if americium is, as would appear, the first lanthanide-type metal in the series, then the $M^{3+}_{(aq)}/M^0_{(s)}$ potentials after americium can be expected to vary in the same sense as those in the lanthanide series.

The crystal structures of the metals are given in Table 8; from this it can be seen that polymorphism is common among the earlier actinides, with plutonium having the greatest number, six forms, in the short interval between room temperature and its rather low melting point, 639.5°. In the cases of δ and δ' plutonium, the metallic radius is 164.4 pm, as compared with 159 pm for the other allotropes. It has been suggested[13] that the higher metallic radius requires a somewhat greater occupation of the $5f$ orbitals than occurs in the other modifications, and that the transition from δ' to ϵ plutonium involves the promotion of an average of 0.4 electrons from the $5f$ to the $6d$ level, an effect which might also explain the anomalous thermal behaviour to be described later.

The arrangement of the metal atoms in the various allotropes of uranium, neptunium and plutonium is difficult to rationalise, whereas the structures of the other actinide metals are relatively straightforward. Thus americium, curium and berkelium all have the lanthanide-type of hexagonal, close-packed structure in which each metal atom has 12 neighbours, and actinium also adopts the 12 coordinate arrangement of β-lanthanum. Both modifications of thorium are based on the cube; α-thorium has the face-centred (or cubic close-packed) arrangement in which each thorium atom is surrounded by 12 others, and β-thorium has the body-centred arrangement in which the coordination number is 8. In protactinium metal, however, each atom has ten almost equidistant neighbours.

The α-form of uranium consists of metal atoms which have only 4 nearest neighbours, two lying at the polar, and two at the equatorial apices of a trigonal bipyramid, while in β-uranium there appear to be no less than 6 different sites, corresponding to "metal valencies" of 3, 4, 5 and 6. This is in marked contrast to α-uranium, for which the calculated "metal valency" is 4, and to γ-uranium, where the corresponding value[9] is 3 and the coordination number is 8. The packing in the neptunium allotropes differs somewhat from that found for uranium: in α-neptunium each metal atom also has only four nearest neighbours, but one half of the neptunium atoms are arranged in the same way as α-uranium, while for the other half the arrangement consists of four atoms lying at

TABLE 8

CRYSTALLOGRAPHIC DATA FOR THE ACTINIDE METALS

Element	Crystal symmetry, temperature range of the phase and space group or structure type	Lattice parameters, Å			Density g/cc (calcd)	Temp.	Ref.
		a_O	b_O	c_O			
Ac	Face-centred cubic, β-La	5.311	–	–	10.07	~ 20	3
Th	α, Face-centred cubic (< 1400°)	5.079	–	–	11.72	~ 20	4
	β, Body-centred cubic (1400–1750°)	4.11	–	–			5
Pa	Body-centred tetragonal	3.929	–	3.241	15.37	~ 20	6
U	α, Orthorhombic (< 667°) D_{2h}^{17}–$Cmcm$	2.854	5.869	4.955	19.07	~ 20	7,8
	β, Tetragonal (667–775°) C_{4v}^{a}–$P4_2nm$	10.75	–	5.65	18.11	700	9,10
	γ, Body-centred cubic (> 775°)	3.534	–	–	18.06	800	11
Np	α, Orthorhombic (< 280°), $Pmcn$ or $Pmna$	4.723	4.887	6.663	20.45	20	1a
	β, Tetragonal (280–577°), $P42_1$	4.897	–	3.388	19.36	313	1b
	γ, Body-centred cubic (> 577°)	3.52	–	–	18.00	600	1b

(Table 8, *continued*)

Pu	α, Monoclinic (< 117°), $P2_1/m$	6.183	4.822 ($\beta = 101.79°$)	10.963	19.82	~ 20	12
	β, Body-centred monoclinic (117–185°) $I2/m$	9.284	10.463 ($\beta = 92.13°$)	7.859	17.70	190	13
	γ, Orthorhombic (185–310°) $Fddd$	3.159	5.768	10.162	17.14	235	14
	δ, Face-centred cubic (310–458°)	4.637	–	–	15.92	320	15
	δ′, Body-centred tetragonal (458–480°)	3.339	–	4.446	16.01	477	16
	ε, Body-centred cubic (> 480°)	3.636	–	–	16.48	490	15
Am	α, Double hexagonal close packed (< 600°), $P6_3/mmc$, α-La	3.4681	–	11.241	13.671	~20	17
	β, cubic (> 600°)	?					17
Cm	Double hexagonal close packed, α-La	3.496	–	11.331	13.51	~20	2
Bk	α, Double hexagonal close packed, α-La	3.416	–	11.069	14.78	–	26
	β, face-centred cubic, $Fm3m\text{-}O_h^5$	4.997	–	–	13.25	–	26

one polar and three equatorial apices of a trigonal bipyramid. In β-neptunium there are again only four nearest neighbours for each metal atom, one-half of the atoms being at the apices of a deformed tetrahedron and the other half at the apices of a flattened tetragonal pyramid. However, the arrangement in γ-neptunium is the same as in γ-uranium.

The packing in three of the six modifications of plutonium is more complicated; in the α-form the atoms are arranged in highly distorted pseudo-hexagonal layers normal to the *b* axis, each atom having 6 neighbours situated in the reflection planes and from 6 to 10 neighbours in adjacent layers giving rise to coordination numbers from 12 to 16. In all, there are 8 different sites: one which has 5 near neighbours at the apices of a trigonal pyramid and 7 more distant neighbours, another which has 3 near neighbours at the corners of an equilateral triangle and 13 more distant neighbours, and the remaining sites have 4 near and 10 more distant neighbours[12]. In the β-form there

References p. 56

are 7 different sites[13] with coordination numbers of 12, 13 or 14.
γ-Plutonium has an interesting 10-coordinate arrangement[14] in which
6 neighbours are at the corners of a hexagon, the remaining four being
approximately tetrahedrally arranged, with two atoms above and two
below the hexagon. There does not appear to be any simple geometrical
relationship between the α, β and γ structures, the last of which
represents almost the closest packing which can be attained for atoms
with a coordination number of 10, but in the remaining modifications
the structures are somewhat simpler. The coordination number in
δ-plutonium is 12, and δ′-plutonium is a slightly distorted form of the
δ-modification, while in ε-plutonium there are 8 nearest neighbours and
6 rather more distant neighbours.

2. PHYSICAL PROPERTIES

Some physical properties of the metals are given in Table 9; it can
be seen from this that their melting points, up to and including americium, are not correlated with those of the corresponding lanthanides
and for the three metals thorium, protactinium and uranium, are

TABLE 9
SOME PHYSICAL PROPERTIES OF THE METALS[a]

Metal	M.p., °C	B.p., °C/760 mm	ΔH_{fusion} kcal/g atom	ΔH_{vap} kcal/g atom	Electrical resistivity, Ω-cm
Ac	1050±50[19]	3300[20]	—	—	—
Th	1750	4000	4.6	130	18×10^{-6} (α)
Pa	1560[21]	—	—	106.6[22]	—
U	1132	3818	4.7	106.7	$25-50 \times 10^{-6}$ (α)
Np	640	—	—	94.3[22]	—
Pu	639.5	3327[23]	—	82.2[23]	145×10^{-6} (α)[b]
Am	994[17]	—	—	~ 55[22]	—
Cm	1340[2]	—	—	—	—
Bk	986[27]	—	—	—	—

[a] Unless otherwise stated, data are from *The Chemistry of the Actinide Elements*,
J.J. Katz and G.T. Seaborg, Methuen, London, 1957.
[b] β-Pu, 110; γ-Pu, 110.5; δ-Pu, 103; δ′-Pu, 105; ε-Pu, 114×10^{-6}.

appreciably lower than the melting points of the d-transition metals hafnium (2150°), tantalum (2996°) and tungsten (3410°). Similarly, the boiling points of the actinide elements are also much lower than those of the d-transition metals mentioned above. However, the melting point of curium approaches that of gadolinium (1312°), its lanthanide analogue, and it would be of interest to have further physical data for this metal, and for the trans-curium metals, for the most recent work indicates that berkelium[27] and californium[28] metals have much lower melting points and are much more volatile than curium.

There are too few other physical data available for the actinide metals other than thorium, uranium and plutonium to allow any comparison to be made with the d-transition and lanthanide elements which is meaningful. However, it is worth noting that the electrical resistivity of both thorium and α-uranium is of the same order as that of iron (9.71 × 10^{-6} Ω/cm), thorium having a lower resistivity than either hafnium or zinc whereas that of α-uranium is about six times the value for tungsten or molybdenum. The electrical resistivities of the allotropes of plutonium are all about 20 times that of tungsten, the value for α-plutonium being higher than that of any other metal at room temperature. The variation of resistivity with temperature among the plutonium allotropes is also exceedingly unusual in that α-, β-, γ- and ϵ-plutonium all have a negative temperature coefficient of resistivity, whereas metals normally have a positive coefficient; nevertheless these four allotropes have positive coefficients of thermal expansion, which is normal for metals. On the other hand, δ- and δ'-plutonium have the normal positive temperature coefficients of resistivity but both have negative coefficients of thermal expansion, so that there is no plutonium allotrope which behaves in the normal metallic manner in respect of both resistivity and thermal expansion.

Zachariasen[13], as mentioned earlier, has suggested that the anomalous metallic radius of plutonium in the δ- and δ'-modifications may be associated with the demotion of about 0.4 electrons from the "valence" shell to the $5f$ shell; it also appears that the negative thermal expansion of the δ phase could be accounted for if the number of $5f$ electrons present decreased from 3.4 at 320° to 3.0 at 465°.

References p. 56

3. CHEMICAL PROPERTIES

The actinide metals have a silvery or greyish silver appearance when a freshly cut surface is exposed but this tarnishes rapidly in air owing to oxidation. At room temperature the oxidation does not proceed beyond the formation of a superficial film on massive pieces of the metals but at elevated temperatures the rate of attack increases and oxidation of the mass becomes complete. Water vapour also reacts with the metals, forming the oxide and hydrogen; the latter can react with the metal to form a hydride, aiding the disintegration of the metal mass, and these reactions are particularly fast with boiling water or steam. A typical reaction with uranium is:

$$7\,U + 6\,H_2O\,(g) \rightarrow 3\,UO_2 + 4\,UH_3$$

Because reactions of this type are, in the case of uranium, more vigorous than the reaction of that metal with oxygen, it is essential to protect actinide metal fuel elements in nuclear reactors from the water coolant or moderator, and it is even more important to ensure that water is never used in attempts to extinguish fires in which actinide metals are involved. This last consideration also applies to carbon dioxide and carbon tetrachloride fire extinguishers, for both compounds react vigorously with the heated actinide metals.

When in a finely divided state the actinide metals oxidise so rapidly on exposure to air at room temperature as to be pyrophoric and in this state the metals can only be handled in an inert gas atmosphere; argon is most commonly used, although helium, where available, has also found some application.

The metals react with a wide variety of non-metals. The compounds formed are generally refractory in character and are of some importance as fuels in high temperature nuclear reactors; examples are given in Table 10. Of these compounds, the hydrides are the least stable to heat and since they are conveniently made simply by heating the massive metal in an atmosphere of hydrogen, their thermal decomposition under vacuum provides a convenient route from massive metal to metal powder. The other compounds with non-metals are made by heating to-

TABLE 10

SOME COMPOUNDS OF THE ACTINIDES WITH NON-METALS[a]

	Stoicheiometry	Known for metal (M)
Hydrides	MH_2	Th, Pu, Am
	M_3H_8	Pu, Am
	MH_3	Pa, U, Pu
	M_4H_{15}	Th, Np
Carbides	MC	Pa, Th, U, Np, Pu
	M_2C_3	Th, U, Np, Pu
	MC_2	Th, U, Np
Silicides[b]	MSi	Th, U, Pu
	MSi_2	Th, U, Np, Pu

[a] Structural data for many of these compounds have been summarised in *The Chemistry of the Actinide Elements*, J.J. Katz and G.T. Seaborg, Methuen, London, 1957.

[b] A variety of other stoicheiometries are known for the uranium- and plutonium-silicon systems.

gether either the constituent elements or the hydride and the non-metal component. The hydrides are extremely reactive towards air or moisture, but the other compounds listed in Table 10 are much more stable in this respect. The actinide metals also form very many identifiable compounds with other metals and some of the alloys with other metals are used for nuclear reactor fuels.

The massive metals are attacked only slowly by mineral acids other than hydrochloric (and, presumably, hydrobromic and hydriodic) acid. Concentrated nitric acid appears to passify thorium and uranium, but addition of fluoride ion seems to remove the passivity. Hydrochloric acid reacts vigorously with the actinide metals, but the dissolution of thorium, protactinium and uranium is incomplete. In the case of protactinium this appears to be the result of hydrolysis to a white, basic compound, while the residue remaining from the reaction of thorium or uranium is black. In both cases the black material is believed to be a hydride hydroxide, the thorium compound[24,25] having the composition $HThO(OH)$ and the uranium compound[24] being $UH(OH)_2$. The latter is said to be pyrophoric when dry.

References p. 56

4. PREPARATION OF THE METALS

Electrolytic or chemical reduction of actinide ions in aqueous
solution is impracticable (See Table 6, p. 26, for the $M^{3+}_{(aq)}/M^0_{(s)}$
potentials) but electrolytic reduction in fused salt media, although never
used, should be possible. The normal method of obtaining the metals is
by reduction of the anhydrous fluorides with magnesium, calcium or
barium at temperatures sufficiently high to melt the metal produced by
the chemical reduction into a suitable ingot or bead. Although the
reduction of anhydrous chlorides or bromides is equally effective, the
fluorides are preferred because they are not hygroscopic whereas the
other halides are. The actinide oxides are also reduced by the alkaline
earth metals but this route does not seem to be used commercially,
probably because the metal produced in this way is generally in the form
of a powder which, as we have seen, oxidises very readily.

5. REFERENCES

1   W.H. Zachariasen, *Acta Cryst.*, (a) 5 (1952) 660; (b) 5 (1952) 664.
2   B.B. Cunningham and J.C. Wallmann, *J. Inorg. Nucl. Chem.*, 26 (1964) 271.
3   J.D. Farr, A.L. Giorgi, M.G. Bowman and R.K. Money, *J. Inorg. Nucl. Chem.*,
    18 (1961) 42.
4   W.J. James and M.E. Straumanis, *Acta Cryst.*, 9 (1956) 376.
5   P. Chiotti, *J. Electrochem. Soc.*, 101 (1954) 567.
6   W.H. Zachariasen, *Acta Cryst.*, 5 (1952) 19; 12 (1959) 698.
7   C.W. Jacob and B.E. Warren, *J. Am. Chem. Soc.*, 59 (1937) 2588.
8   E.F. Sturcken and B. Post, *Acta Cryst.*, 13 (1960) 852.
9   J. Thewlis and H. Steeple, *Acta Cryst.*, 7 (1954) 323.
10 H. Steeple and T. Ashworth, *Acta Cryst.*, 21 (1966) 995.
11 H.H. Klepper and P. Chiotti, *ISC Report 893* (1957).
12 W.H. Zachariasen and F.H. Ellinger, *Acta Cryst.*, 16 (1963) 777.
13 W.H. Zachariasen and F.H. Ellinger, *Acta Cryst.*, 16 (1963) 369.
14 W.H. Zachariasen and F.H. Ellinger, *Acta Cryst.*, 8 (1955) 431.
15 E.R. Jette, *J. Chem. Phys.*, 23 (1955) 365.
16 F.H. Ellinger, *(J. Metals) Trans. AIME*, 8 (1965) 1256.
17 D.B. McWhan, B.B. Cunningham and J.C. Wallmann, *J. Inorg. Nucl. Chem.*,
    24 (1962) 1025.
18 W.H. Zachariasen, *U.S. At. Energy Comm. Unclassified Document TID-7683*
    (1963).
19 J.G. Stites, M.L. Salutsky and B.D. Stone, *J. Am. Chem. Soc.*, 77(1949) 237.
20 K.W. Foster, *U.S. At. Energy Comm. Unclassified Document MLM-901* (1953).

21   B.B. Cunningham, *Proc. Colloq. sur la Physicochim. Protactinium*, Orsay, 1965; C.N.R.S., Paris, 1966, p. 45.

22   R.J. Ackermann and R.J. Thorn, in *Thermodynamics*, Proc. Vienna Symp., 1965; I.A.E.A., Vienna, 1 (1966) 243.

23   R.N.R. Mulford, in *Thermodynamics*, Proc. Vienna Symp., 1965; I.A.E.A., Vienna, 1 (1966) 231.

24   A.G. Karabash, *Zhur. Neorgan. Khim.*, 3 (1958) 986.

25   L.I. Katzin, L. Kaplan and T. Steitz, *Inorg. Chem.*, 1 (1962) 963.

26   J.R. Peterson, J.A. Fahey and R.D. Baybarz, *J. Inorg. Nucl. Chem.*, 33 (1971) 3345.

27   J.A. Fahey, J.R. Peterson and R.D. Baybarz, *Inorg. Nucl. Chem. Letters*, 8 (1972) 101.

28   T.C. Parsons, personal communication.

Chapter 5

COMPOUNDS WITH MAIN GROUP V ELEMENTS

1. INTRODUCTION

Compounds of the actinides with nitrogen, phosphorus, arsenic, antimony and bismuth, often collectively but inelegantly referred to as *pnictides*, have been extensively investigated during the past decade because of their possible use as nuclear fuel materials; in this role the nitrides could be of particular importance because of their high melting points. Reviews covering these compounds have appeared in textbooks on the chemistry of uranium[1] and plutonium[2], and in a book reviewing progress in lanthanide and actinide chemistry[3].

2. PREPARATIVE METHODS

These compounds are usually prepared by heating the actinide metal either with the Group V element in a sealed tube or with its hydride (*e.g.* PH_3^4, AsH_3^5) when this is available. Preparative reactions of this type go to completion only when the actinide metal is in a finely divided form, for in the massive form the metal may become coated with a protective layer of the product which inhibits further reaction. The finely divided actinide metal is usually obtained (p. 54) by heating the massive metal in hydrogen to form the hydride and subsequently decomposing this in a vacuum at higher temperature.

The experimental conditions are generally arranged to yield compounds in which the Group V element content is a maximum. When necessary, such products are degraded to compounds with a lower proportion of the Group V element by heating them in a vacuum or in a stream of an inert gas, such as argon; the liberated Group V element distils away leaving the product of the dissociation. Alternatively, the primary material can be reduced by heating it with the appropriate amount of the finely divided actinide metal itself, or its hydride. It is also possible to produce actinide nitrides or phosphides by heating a mixture of the actinide oxide with carbon in the presence of nitrogen or phosphorus, but the compound produced is often contaminated with carbon or the actinide oxide, and it may also be non-stoicheiometric.

3. NITRIDES

The identified nitrides are Th_3N_4, U_2N_3 (two crystal modifications are known for each of these) and the mononitrides, MN (M = Th, Pa, U, Np, Pu, Am); a compound, UN_2, was once thought to exist but it cannot, in fact, be prepared. Apart from PaN, for which crystallographic data are lacking, all the mononitrides adopt the rock salt structure and are mutually miscible, as one would expect from the similarity in ionic radii of the actinides involved. Thorium has also been stated to form a nitride Th_2N_3, but the crystallographic data reported for the product thus described are actually those of a nitride oxide[6], Th_2N_2O, which was prepared by heating a mixture of ThN (3 moles) and ThO_2 (1 mole) in nitrogen at 1700°. A greenish yellow nitride imide, $Th_2N_2(NH)$, analogous to the nitride oxide, is also known, obtained[7] by heating thorium metal in ammonia gas at high (> 2500 atm) pressure. This crystallises with a distorted La_2O_3 type lattice, in which thorium is presumably 7-coordinate, and decomposes above 270° to yield βTh_3N_4. This is a brown solid which reacts with ammonia under pressure (700 atm) at 360° to reform the nitride imide[7].

The structure of αTh_3N_4, made by heating ThN with the stoicheiometric quantity of nitrogen at 1320°, is closely related to that of the nitride oxide[6]. αTh_3N_4, a maroon solid, is also obtained by heating finely divided thorium metal in nitrogen at 500–1000°. The compound can be regarded as $(Th^{4+})_3(N^{3-})_4$ and is, as would be expected, diamagnetic; it decomposes to the greenish-yellow mononitride above 1400° in a vacuum. These thorium compounds are somewhat hygroscopic; they are decomposed by water with the formation of ammonia. Th_3N_4 has a high electrical resistivity (10^5–10^8 Ω cm at $20°$) and the temperature coefficient of resistivity is typical of an ionic semiconductor. In contrast, the electrical resistivity of ThN is only 2×10^{-5} Ω cm at $20°$, almost the same as that of thorium metal. The formal charge on the nitride ion is believed[13] to be less than 3, and the charge distribution can be represented as Th^{4+} $N^{2·53-}$ ($1.47e^-$).

Apart from UN, the mononitrides of the actinides have scarcely been investigated. The only known protactinium nitride is formed by heating protactinium tetrachloride or pentachloride in ammonia gas at $800°$; the product is a bright yellow solid[8]. Although most reviews have accepted this product to be PaN, more work on this system is necessary, particularly as the indubitable mononitrides are all dark, metal-like solids.

The dark grey uranium compound, UN, is usually obtained by heating αU_2N_{3+x} (see below) above $1100°$ in a vacuum or in a stream of argon. It is pyrophoric in air when finely divided, and because of its high melting point ($2850°$) it has attracted much research interest as a possible nuclear reactor fuel material. It is stable in water at $100°$ but yields UO_2, ammonia, hydrogen, and some U_2N_{3+x} with water vapour at high temperatures. The uranium nitride systems have been extensively reviewed by Cordfunke[1]. Neptunium mononitride is obtained by heating the metal in nitrogen at $600°$ in the presence of a small amount of hydrogen (0.5 %) which serves to catalyse the reaction by breaking down the metal to a powder by way of the hydride[9]; the plutonium compound is formed[10] by heating plutonium hydride in nitrogen at $400°$ and the americium compound[11] by heating americium hydride in ammonia at $800°$. A product, also described as AmN, obtained by heating americium[11] metal in nitrogen at $750°$ was probably a mixture of the monoxide and mononitride[3]. There is no other information available on the protactinium, neptunium and americium nitride systems. Information about the chemistry of these mononitrides is very limited, but PuN is known to hydrolyse to PuO_2, ammonia and hydrogen in water vapour; hydrated plutonium dioxide is formed when the hydrolysis takes place below $350°$.

In the case of uranium, the nitride phase obtained when the powdered metal is heated in nitrogen or ammonia above $400°$ is $\alpha U_2 N_{3+x}$, a product in which the composition varies from $UN_{1.60}$ to $UN_{1.75}$ depending upon the conditions of preparation. The crystal structure is the same as that of the C-type lanthanide oxides; in the latter the metals are 6-coordinate (see Chapter 6, p. 66). The structure of $\beta U_2 N_3$, however, is the $La_2 O_3$ type (A-type lanthanide oxide) and uranium is presumably 7-coordinate in this compound, with the coordination geometry of a capped octahedron. $\beta U_2 N_3$ is obtained by quenching $\alpha U_2 N_3$ from $1100°$ and its composition range[12] is rather narrow, $UN_{1.44-1.45}$. The behaviour of $\alpha U_2 N_3$ towards water vapour is similar to that of the mononitride.

4. PHOSPHIDES, ARSENIDES, ANTIMONIDES AND BISMUTHIDES

The thorium and uranium systems with the Group V elements giving these compounds have been investigated quite extensively; the plutonium systems are much less well known and the neptunium and americium systems are almost unknown, for only a single compound of each actinide metal has been reported. The types of compound formed in these systems are shown in Table 11, from which it can be seen that the known compounds fall into three isostructural series.

TABLE 11

ACTINIDE PHOSPHIDES, ARSENIDES, ANTIMONIDES AND BISMUTHIDES

Stoicheiometry	MA	$M_3 A_4$	MA_2
Thorium	$A = P^a$, As,Sb,(Bi?)[b]	A = P,As,Sb,Bi	A = As,Sb,Bi
Uranium	A = P,As,Sb,Bi	A = P,As,Sb,Bi	A = P,As,Sb,Bi
Neptunium	A = Sb	—	—
Plutonium	A = P,As,Sb,Bi	—	—
Americium	A = Sb	—	—
Structure type	fcc,NaCl[c]	bcc,$Th_3 P_4^d$	tet.,$Fe_2 As^e$

[a]Substoicheiometric, ThP_{1-x}.
[b]Existence doubtful; products are not NaCl or CsCl type structures[14].
[c]fcc, face-centred cubic.
[d]bcc, body-centred cubic.
[e]tet., tetragonal.

The thorium[14,15] and uranium[16-18] compounds of the type MA_2 are usually prepared by heating the stoicheiometric quantities of the elements together in a sealed tube or ($ThAs_2$) by reaction of the metal with arsine. The reaction of phosphine with thorium yields only Th_3P_4, whereas uranium metal reacts with phosphorus or with phosphine at $500°$ to yield UP_2. There is X-ray evidence for the formation of a small amount of a tetragonal phase in the preparations of Th_3P_4 which may indicate that some thorium diphosphide is present[19], but the latter compound is not known to have been isolated as a pure phase.

Compounds of the type MA_2 presumably involve a polyanionic structure, possibly of the type M^{4+} $(A\text{-}A)^{4-}$. When heated with the stoicheiometric quantity of thorium powder in a vacuum, the thorium compounds yield M_3A_4 and MA; under such reaction conditions Th_3P_4 yields the dark blue, sub-stoicheiometric monophosphide, ThP_{1-x}. U_3P_4 is obtained directly from UP_2 by thermal decomposition in argon at $800°$, and at higher temperatures (1300-$1400°$) the grey monophosphide, UP, is produced. The composition of this last is very close to stoicheiometric; the compound is very resistant to hydrolysis.

The arsenides UAs_2 and U_3As_4 are prepared by heating stoicheiometric quantities of the elements together at $800°$; when the temperature is raised to 1300-$1400°$ in a vacuum dissociation to UAs occurs. This last compound has also been made[5] by heating uranium with the stoicheiometric quantity of arsine at $300°$, followed by heating of the product in a vacuum at 1300-$1400°$. The various antimonides and bismuthides are usually prepared by heating together stoicheiometric quantities of the elements in a sealed tube at 800-$900°$. The neptunium, plutonium and americium mono-compounds, MA, are all obtained in a similar manner; structural and magnetic data for these compounds have been discussed in detail in a review[3].

5. REFERENCES

1 E.H.P. Cordfunke, *The Chemistry of Uranium*, Elsevier, Amsterdam, 1969, Chapter 11.
2 J.M. Cleveland, *The Chemistry of Plutonium*, Gordon and Breach, New York, 1970, Chapter 14.
3 R.M. Dell and N.J. Bridger, *Intern. Rev. Sci.*, 7 (1972) 211; Medical and Technical Publishing Co., and Butterworth, London.
4 M. Allbutt, A.R. Junkison and R.F. Carney, *Proc. Brit. Ceram. Soc.*, 7 (1967) 111.
5 Y. Baskin, *J. Inorg. Nucl. Chem.*, 29 (1967) 2480.
6 R. Benz and W.H. Zachariasen, *Acta Cryst.*, 21 (1966) 838.
7 R. Juza and H. Gerke, *Z. Anorg. Allgem. Chem.*, 363 (1968) 245.
8 P.A. Sellars, S. Fried, R.E. Elson and W.H. Zachariasen, *J. Am. Chem. Soc.*, 76 (1954) 5935.
9 W.M. Olsen and R.N.R. Mulford, *J. Phys. Chem.*, 70 (1966) 2932.
10 M. Allbutt and R.M. Dell, *J. Nucl. Materials*, 24 (1967) 1.
11 Y. Akimoto, *J. Inorg. Nucl. Chem.*, 29 (1967) 2650.
12 Y. Sasa and T. Atoda, *J. Am. Ceram. Soc.*, 53 (1970) 102.
13 A.B. Auskern and S. Aronson, *J. Phys. Chem. Solids*, 28 (1967) 1069.
14 R. Ferro, *Acta Cryst.*, 10 (1957) 476.
15 R. Ferro, *Acta Cryst.*, 8 (1955) 360; 9 (1956) 817.
16 Y. Baskin, *J. Am. Ceram. Soc.*, 49 (1966) 541.
17 W. Trzebiatowski, A. Sepichowska and A. Zygmunt, *Bull. Acad. Polon. Sci., Ser. Sci. Chim.*, 12 (1964) 687.
18 W. Trzebiatowski and A. Zygmunt, *Bull. Acad. Polon. Sci., Ser. Sci. Chim.*, 14 (1966) 495.
19 F. Hulliger, *Nature*, 209 (1966) 499.

Chapter 6

OXIDES, HYDROXIDES, SULPHIDES, SELENIDES AND TELLURIDES

1. INTRODUCTION

The actinide oxides are of importance in the atomic energy industry because their refractory nature makes them suitable for the preparation of ceramic fuel elements for use in high-temperature nuclear reactors. For example, uranium dioxide melts at 2800°. These compounds have therefore been extensively investigated; although the properties of the other chalcogenides are rather less well-known, some of them may also prove useful in the same connection. Detailed reviews of the uranium[1] and plutonium[2] compounds are available, and general reviews of the actinide oxides[3], mixed oxide systems[4] and sulphides, selenides and tellurides[5] cover the field in much greater depth than is attempted here. However, actinide polonides do not seem to have been reported, although the lanthanide polonides are quite well known.

2. OXIDES

The principal actinide oxide phases are given in Table 12. At first sight, the oxide systems appear to be relatively simple, but the systems are, in fact, extremely complicated because of the many, and in some cases wide, deviations from stoicheiometry which occur. For example, the lattice of uranium dioxide, UO_2, can both accommodate oxygen interstitially, leading to super-stoicheiometric phases UO_{2+x}, and also lose oxygen, to form sub-stoicheiometric phases UO_{2-x}. Behaviour of this kind has been discussed very fully in the reviews cited above and this topic is not considered further here.

From Table 12 it is apparent that the *anhydrous* oxides in which

TABLE 12

PRINCIPAL ACTINIDE OXIDE PHASES (IDEAL FORMULAE)

	ThO	PaO	UO	NpO	PuO	AmO		BkO	
Ac_2O_3	–	–	–	–	Pu_2O_3	Am_2O_3	Cm_2O_3	Bk_2O_3	Cf_2O_3
	ThO_2	PaO_2	UO_2	NpO_2	PuO_2	AmO_2	CmO_2	BkO_2	CfO_2
			U_4O_9						
		$PaO_{2.3}$	$UO_{2.3}$						
			U_3O_8	Np_3O_8					
		Pa_2O_5	U_2O_5	Np_2O_5					
			UO_3	$(NpO_3 \cdot H_2O)(PuO_3 \cdot O \cdot 8\ H_2O)$					

the metals exhibit oxidation states greater than +4 are known only for
protactinium, uranium and neptunium; this demonstrates the decreasing
thermodynamic stability of the higher oxidation states as the atomic
number increases. Evidently the metals in their highest possible
oxidation states have become sufficiently oxidising to oxidise the O^{2-}
ion to molecular oxygen, because the energy contribution from other
factors, such as the lattice energy, is too small for the compounds
formed to be thermodynamically stable.

The known actinide monoxide phases all have face-centred cubic
(NaCl type) symmetry; they are semi-metallic in character. Even AmO,
a grey solid with a metallic lustre which is prepared[6] by the reaction of
americium metal with the stoicheiometric quantity of oxygen at 850°,
is semi-metallic despite an expected analogy between americium and
europium. There is, therefore, no example of an actinide monoxide
which contains the bipositive, actinide metal ion, although the trace
level chemical behaviour of nobelium in aqueous solution suggests that
the bipositive state is particularly stable with respect to oxidation for
this element, so that it may well form an ionic monoxide. Plutonium
monoxide can be made in the same way as the americium compound,
while evidence for monoxides of the other elements usually rests on
X-ray identifications of oxide films formed on the surface of specimens
of the metals[7].

There have been several unsuccessful attempts to produce these
monoxides by reducing the dioxide with the appropriate metal at high
temperatures and, in the case of UO, by heating UC with UO_2 to above

$1600°$. In every case, the UO phase produced by such reactions contained carbon or nitrogen; an example, quoted by Roberts[3], of the product of the last reaction had the composition $UC_{0.57}O_{0.43}$. Subsequent work indicated that the nucleation of the uranium monoxide phase is catalysed by nitride or carbide[8].

Sesquioxides, M_2O_3, are known only for actinium and the elements plutonium to californium inclusive; the plutonium, americium and berkelium compounds are obtained by reducing the dioxides. The formation of Pu_2O_3 requires reaction of the dioxide with plutonium metal at $1500°$, whereas the americium and berkelium compounds are obtained by reducing the dioxides in hydrogen at $600°$; on the other hand, Cm_2O_3 is prepared by thermal decomposition of curium(III) oxalate, the product then being heated at $600°$ in a vacuum. The californium compound has been prepared on the microgram scale by sorbing Cf^{3+} ions on beads of a cation-exchange resin and then igniting the resin at $1200°$; the product is then heated at $600°$ in hydrogen.

The sesquioxides from plutonium to californium inclusive adopt the lanthanide oxide C-type structure (body-centred cubic, Mn_2O_3) in which the metal atom lies at the centre of a cube and oxygen atoms occupy six of.the eight corners of the cube, the vacant corners being either on a face diagonal of the cube or on a body diagonal through the cube; however, the plutonium oxide with this structure is only obtained by reaction of the dioxide with carbon at $1625°$ or when the dioxide is heated above $1650°$ in a vacuum, and the stoicheiometry seems to vary from $PuO_{1.5}$ to $PuO_{1.6}$. The actinium compound, Ac_2O_3, the stoicheiometric Pu_2O_3 made by reducing the dioxide with the metal at $1500°$, and a second form of Am_2O_3 produced by reduction of the dioxide with hydrogen at $800°$, all adopt the lanthanide A-type structure (hexagonal symmetry). In this, seven oxygen atoms are arranged around the metal atom in the form of a singly capped octahedron, an arrangement which applies particularly to cases where the ligand–ligand (in this case oxygen) separation is large. Only one example of the B-type lanthanide oxide phase (monoclinic, Sm_2O_3) has been recorded for the actinides; this is a second[9], and presumably high temperature, form of Cf_2O_3. There are three different structural sites in crystals of the B-type oxides; all of them involve the 7-coordinate metal, one having the A-type capped octahedral geometry, the other two being singly capped trigonal prismatic arrangements which differ in their metal-oxygen distances.

All the known actinide dioxides are of face-centred cubic symmetry (fluorite type); they are obtained by air ignition of compounds such as the oxalates or hydroxides, a procedure which, with protactinium or uranium, yields higher oxides, Pa_2O_5 and U_3O_8 or UO_3; however, curium dioxide is formed only when the oxalate or hydroxide is heated in oxygen. Protactinium dioxide is prepared by reduction of the pentoxide in hydrogen at $1500°$, and uranium dioxide results from reduction of the higher oxides in hydrogen at $300-800°$ or by electrolytic reduction[10] of uranyl chloride in a fused KCl/LiCl eutectic at $450-500°$. Uranium dioxide is readily oxidised and, when finely divided, it is pyrophoric in air.

Oxides intermediate in composition between MO_2 and M_2O_5 are known for both protactinium and uranium; the tetragonal oxide $PaO_{2.33}(Pa_3O_7)$ appears to be the main phase formed when PaO_2 is heated in oxygen at $800°$ and this phase has also been observed as an intermediate in the reduction of protactinium pentoxide in hydrogen or under vacuum at high temperatures[11]. The oxidation of UO_2 at temperatures between $135°$ and $350°$ leads to oxide phases of similar composition, $UO_{2.3-2.4}$, for which three crystal modifications have been reported, but oxidation of UO_2 in air under reduced pressure (10^{-2} torr) at $900°$ yields U_4O_9, a cubic phase of large cell dimensions[12], which also exists in three different crystallographic forms. There does not appear to be a protactinium oxide analogous to this uranium phase. Although neptunium forms two well defined higher oxides, Np_3O_8 and Np_2O_5, there is no evidence at present for phases intermediate between these compounds and NpO_2; both NpO_2 and PuO_2 are, in fact, quite inert in oxygen, in contrast to UO_2.

Triuranium octaoxide, U_3O_8, is the product formed when uranium(IV) oxalates, or similar uranyl salts, are ignited in air at temperatures below $900°$. This oxide exists in three crystallographic modifications, the α-form being the uranium oxide phase which is stable at temperatures above $500°$ when in contact with oxygen. The β-phase is formed when the α-phase is heated in oxygen (1 atm) at $1350°$ and slowly cooled, but it reverts to the α-phase when heated at $125°$; at present no explanation has been given for this behaviour. The structures of the two phases are closely related, the α-form being of c-face-centred orthorhombic symmetry ($a = 672$, $b = 1196$ and $c = 415$ pm) whilst the β-form is also orthorhombic[13] ($a = 705$, $b = 1142$ and $c = 829$ pm), so that the principal difference between the two phases lies in the doubling of the c parameter in the β-form. There is probably very little difference in energy between the α- and β- forms and Cordfunke[1] suggests that the transformation behaviour is probably of kinetic origin.

The corresponding neptunium oxide, Np_3O_8, is prepared by heating neptunium(V) or neptunium(VI) compounds in air at $275-450°$ or by heating the hydroxide in nitrogen dioxide at $300-450°$. This also has an orthorhombic cell[14] ($a = 657.9$, $b = 408$ and $c = 418.2$ pm).

Pentoxides, M_2O_5, are known for protactinium, uranium and neptunium; the white protactinium compound results when protactinium(V) hydroxide is ignited in air, whereas the violet uranium compound[15] is obtained by reduction of U_3O_8 or UO_3 by hydrogen sulphide at relatively low temperatures, U_3O_8 between $0°$ and $140°$, α UO_3 at $80°$ and amorphous UO_3 at $0°$. The sulphur liberated during the reduction is removed by extracting it into carbon disulphide. The product obtained by the reaction with amorphous UO_3 is pyrophoric in air yielding UO_2, presumably by way of reaction with free sulphur. It is also obtained by the reduction of U_3O_8 by aqueous uranium(IV) sulphate at $22°$:

$$2\,U_3O_8(s) + U^{4+}_{(aq)} + H_2O \longrightarrow 3\,U_2O_5(s) + UO^{2+}_{2(aq)} + 2\,H^+_{(aq)}$$

The U_2O_5 phase possesses hexagonal symmetry ($a = 399.6$, $c = 411.7$ pm) and it disproportionates into U_3O_8 and U_4O_9 at $250°$; it is oxidised in air, yielding U_3O_8 between $300°$ and $400°$. Dark brown Np_2O_5 is obtained in a poorly crystalline form by heating monohydrated neptunium trioxide at $300°$ in a vacuum[16]; a more crystalline product is precipitated when ozone is passed through a solution of neptunium(V) in molten lithium perchlorate or by the reaction of neptunium metal with molten lithium perchlorate[17]. The crystal symmetry is monoclinic ($a = 418.3, b = 658.4, c = 408.6$ pm; $\beta = 90.32°$) and the cell is related to that of orthorhombic Np_3O_8 by changing the lattice parameters a, b, c of the latter to b, c, a in Np_2O_5.

The only anhydrous trioxide known is the uranium compound; this exists in 7 crystal modifications, the common forms being amorphous UO_3, prepared by heating hydrated uranium peroxide at 450°, and γUO_3, obtained by heating $UO_2(NO_3)_2 \cdot 6\,H_2O$ at 400–600°. Many of the seven forms are known to involve oxygen-bridged structures in which the uranyl(VI), O=U=O, unit is present. This group is known to be absent in δUO_3 (cubic, ReO_3) and the structure in this modification presumably consists of linked UO_6 octahedra.

3. PEROXIDES

The addition of hydrogen peroxide to an aqueous solution of protactinium(V) in dilute sulphuric acid yields a white precipitate of a peroxide with the composition[18] $Pa_2O_9 \cdot 3\,H_2O$; the compound is unstable and loses oxygen slowly at room temperature. Uranium peroxide is precipitated as the pale yellow tetrahydrate, $UO_4 \cdot 4\,H_2O$, when hydrogen peroxide is added to an aqueous solution of a uranyl salt; water is lost at 40° to form the dihydrate, and complete decomposition to UO_3, oxygen and water occurs above 225°. There is no doubt that the tetra- and dihydrates are true peroxides[19], based on uranium(VI). Neptunium peroxide, a colourless, flocculent precipitate obtained when hydrogen peroxide is added to a solution of neptunium(IV) in dilute mineral acid, is isomorphous[20] with $UO_4 \cdot 2\,H_2O$ but is apparently based on neptunium(IV), not neptunium(VI). The corresponding plutonium compound, prepared in the same way from aqueous solutions of plutonium(IV), appears to be non-stoicheiometric, the composition and crystal symmetry depending upon the precipitation conditions, including the anions present in the solution; it decomposes above 135° to PuO_2.

A brownish-rose neptunium(V) peroxide of uncertain composition is precipitated when a solution of neptunium(V) in 0.5 M $HClO_4/3 M H_2O_2$ is adjusted to pH 3.5 by the addition of sodium hydroxide. Reduction to neptunium(IV) occurs when this product is dissolved in acid, apparently as the result of the reaction:

$$2 NpO_2^+ + 6 H^+ + H_2O_2 \rightleftharpoons 2 Np^{4+} + O_2 + 4 H_2O$$

Spectrophotometric studies indicate that the red-orange solution obtained by treating aqueous alkaline neptunium(VI) with hydrogen peroxide also contains neptunium(V) peroxide[34].

4. HYDROXIDES

The hydroxides of the terpositive metals have not been extensively investigated; it is known that bases, such as aqueous ammonia, yield precipitates with solutions of the actinides(III), oxidation being almost instantaneous with uranium(III) and neptunium(III), and quite rapid with plutonium(III). The precipitates obtained with americium(III)–californium(III) usually appear gelatinous when first formed, but it has been shown that when the amorphous americium(III) product is allowed to age, it becomes crystalline, and the resulting $Am(OH)_3$ is then isostructural with $Nd(OH)_3$, the crystal having hexagonal symmetry[21] ($a = 642.6, c = 374.5$ pm).

The actinide(IV) hydroxides (Th, Pa, U, Np, Pu) are all very insoluble in water, and are precipitated as gelatinous masses when alkalis are added to aqueous solutions of the quadripositive element. The protactinium compound is rapidly oxidised to protactinium(V), and the uranium compound, initially green, blackens on standing because of oxidation. The neptunium(IV) and plutonium(IV) compounds are more stable in this respect; all these compounds are presumably hydrated forms of the tetrahydroxide, $M(OH)_4$. Apart from solubility data, nothing is known about them. The same applies to protactinium(V), neptunium(V), plutonium(V) and americium(V) hydroxide species; all are precipitated by alkali from solutions of the quinquevalent metal, the plutonium compound being the most difficult to prepare because of its tendency to disproportionate. All of the foregoing hydroxides are basic in character; they dissolve in acids to form salts and are almost completely insoluble in alkalis.

Hydroxides of the sexivalent elements, of the form $MO_2(OH)_2$, have been found for uranium, neptunium and plutonium only. Uranyl hydroxide is known to exist in a number of crystallographically distinct forms. One of these, $\alpha UO_3 \cdot H_2O$, is obtained by heating the dihydrate, $UO_3 \cdot 2 H_2O$ (or $UO_2(OH)_2 \cdot H_2O$), at $80°$ in air. This dihydrate is the usual product of the hydrolysis of uranyl species. A second form, $\beta UO_3 \cdot H_2O$, is prepared by heating anhydrous UO_3 with water at $180°$ or by hydrolysis of uranyl acetate in solution at $110°$ in a sealed tube[22]. The dihydrate, $NpO_3 \cdot 2 H_2O$, and the monohydrate, prepared[16] by oxidising neptunium(IV) hydroxide suspended in water with ozone at $18°$ and $90°$ respectively, show some crystallographic similarity to the UO_3 hydrates, and can be considered as the hydrated, $NpO_2(OH)_2 \cdot H_2O$, and anhydrous, $NpO_2(OH)_2$, neptunyl hydroxides. The plutonium compound, $PuO_3 \cdot 0.8 H_2O$, obtained by ozone oxidation of an aqueous suspension of plutonium(IV) hydroxide at $90°$, is not isostructural with any uranyl hydroxide species[16]. Nevertheless, it is probably best regarded as a hydroxyl ion deficient form of plutonyl hydroxide. The hydrated plutonyl hydroxide, $PuO_2(OH)_2 \cdot H_2O$, is precipitated by ammonia from aqueous solutions of plutonium(VI), a reaction previously thought to yield the diplutonate, $(NH_4)_2Pu_2O_7$. The hydroxides of the sexivalent elements yield actinyl(VI) salts, $MO_2X_2^I$, with acids; with strong alkali, species such as the uranates(VI) appear to be formed. These last are more usually prepared by heating the actinide oxide with an alkali, or alkaline earth, metal oxide. Some of the products produced in this way are discussed in the following section.

The hydrated septavalent neptunium hydroxide, $NpO_2(OH)_3 \cdot n H_2O$, has also been isolated by oxidation of neptunium(VI) to neptunium(VII) with ozone or other suitable oxidants and neutralisation of the resulting solution, whereupon the black hydroxide is precipitated[24]. Derivatives of the hydroxide are discussed below.

References p. 76

5. ALKALI AND ALKALINE EARTH METALLATES

Very few compounds of the tripositive actinides are known, largely because the ease of oxidation of the actinide(III) oxides makes their preparation extremely difficult. Even $LiAmO_2$ (monoclinic, α-$LiEuO_2$ type), prepared by heating a mixture of the two component oxides at 400–800°, is liable to oxidation unless the preparation is carried out in an inert or reducing atmosphere. The compounds $BaPu_2O_4$ and $BaAm_2O_4$, prepared from the oxides at 1250° in a vacuum, appear to be the only other examples of actinide(III) metallates of the Group I and II metals.

The alkali metal derivatives of the actinide dioxides, prepared from the mixed oxides at 600–800°, are of two types, $M_2^I M^{IV} O_3$ (Th, Am) and $Li_8 M^{IV} O_6$(Pu, Am). In the latter the configuration of the actinide M^{4+} ion is octahedral. Compound formation does not appear to occur with BeO, MgO or CaO, the systems forming simple eutectics. However, both strontium and barium perovskite compounds, $M^{II} M^{IV} O_3$, can be prepared. The compounds $BaM^{IV} O_3$ (Th, Pa, U, Np, Pu, Am) are all of cubic symmetry whereas the strontium compounds (Pa, U, Np, Pu) are not. The available X-ray structural data have been discussed by Keller[4].

Three types of alkali metal-actinide(V) compounds have been reported for the elements from protactinium to americium inclusive, $M^I M^V O_3$(K, Rb have cubic perovskite structures and Na has a deformed perovskite structure), $M_3^I M^V O_4$ and $M_7^I M^V O_6$. The principal methods of preparing them are by either the oxidation of an actinide dioxide/alkali metal oxide mixture with oxygen or the thermal decomposition (or reduction) of an actinide(VI) metallate. The alkaline earth systems are less well-known; compounds of the type $M^{II}(UO_3)_2$ and $M_2^{II} U_2 O_7$ have been reported for uranium and the structures of the calcium compounds are thought[25] to include the UO_2^+ ion.

The best-known actinide(IV) systems are those formed by lithium and sodium, particularly for uranium(VI). The main types of compound formed with the alkali metals are $Li_2 U_3 O_{10}$, $M_2^I M_2^{VI} O_7$, $M_2^I M^{VI} O_4$, $M_4^I M^{VI} O_5$ and $M_6^I M^{VI} O_6$. They are prepared by heating an actinide oxide ($U_3 O_8$, MO_2(Np, Pu, Am)) with an alkali metal oxide, hydroxide, peroxide or carbonate at 400–1300° in oxygen. Alkali metal oxides are usually preferred for preparing compounds of high M^I/M^{VI} ratio. In the case of the alkaline earth metals, the common types of compound are of the forms $M^{II}M^{VI}O_4$ (U, Np, Pu), $M_2^{II}UO_5$ and $M_3^{II}M^{VI}O_6$(U, Np, Pu, Am). They are prepared in a similar manner to the alkali metal compounds. Polyuranates (e.g. $Ba_2U_3O_{11}$) are also known. Compounds of septivalent neptunium and plutonium have been reported only recently; the barium compounds, $Ba_3(M^{VII}O_5)_2 \cdot n\ H_2O$, were first obtained[24] by oxidation of neptunium(VI) and plutonium(VI) in alkaline solution with ozone, hypochlorite, hypobromite and a variety of other oxidising agents, followed by addition of a solution of barium hydroxide in sodium hydroxide (0.5 M), whereupon the compounds are precipitated. Subsequently[26] the anhydrous barium compound has been prepared by the reaction of $NpO_3 \cdot H_2O$ with barium peroxide in oxygen at 400-500°. These compounds are evidently derived from the hydroxide, $NpO_2(OH)_3$, discussed above.

The other known neptunium(VII) and plutonium(VII) species are of the form Li_5MO_6; these contain isolated MO_6 octahedra. They are made by reaction of a mixture of lithium oxide and the actinide dioxide in oxygen[27] at 400-420°. However, Li_5NpO_6 and Li_5PuO_6 dissolve in water or aqueous alkali to yield dark green solutions characteristic of the MO_5^{3-} ions[27]. Somewhat similar barium lithium and barium sodium neptunates(VII), $Ba_2 Li$ (or Na)NpO_6, are obtained by heating $NpO_3 \cdot H_2O$ with a mixture of barium and lithium or sodium peroxides at 450°. They are cubic-ordered perovskite compounds[26].

6. SULPHIDES, SELENIDES AND TELLURIDES

The known compounds are listed in Table 13, from which it can be seen that the sulphides and selenides of a given actinide element have the same crystal structure but are not isostructural with the corresponding telluride. In this group of compounds, the sulphides are generally much more stable to heat than the analogous selenides; the tellurides have the lowest thermal stability of all. Most of the compounds are prepared in a similar manner to the nitrides and

TABLE 13

ACTINIDE SULPHIDES, SELENIDES AND TELLURIDES

Stoicheiometry	MA	M_3A_4	M_2A_3	M_3A_5	M_7A_{12}	MA_2	M_2A_5	MA_3
Actinium	–	–	A = S[a]	–	–	–	–	–
Thorium	A = S,Se,Te[b]	–	A = S,Se,Te[c]	–	A = S,Se	A = S,Se,Te[d]	A = S,Se	A = Te
Uranium	A = S,Se,Te	A = Se,Te	A = S,Se,Te[e]	A = S,Te	A = Te	A = S,Se,Te[f]	A = S,Te	A = S,Se,Te
Neptunium	A = S	A = S,(?),Se,Te	A = S[l]	A = S	–	–	A = S	A = S
Plutonium	A = S,Se,Te	A = S,Se	A = S,Se,Te[g,l]	–	–	A = S,Se,Te[h,l]	–	–
Americium	–	A = Se,Te	A = S[e]	–	–	A = S,Se	–	–
Curium	–	–	A = S[e]	–	–	–	–	–
Berkelium	–	–	A = S[e]	–	–	–	–	–
Californium	–	–	A = S[e]	–	–	–	–	–
Structure type	fcc[i],NaCl	bcc[j],Th_3P_4	orth.[k],Sb_2S_3		hexagonal	orth.[k], $PbCl_2$	tetragonal	monoclinic, $ZrSe_3$

[a] Cubic, Ce_2S_3; [b] ThTe, body-centred cubic, CsCl; [c] Th_2Te_3, hexagonal; [d] $ThTe_2$, hexagonal;
[e] U_2Te_3, Am_2S_3, Cm_2S_3, Bk_2S_3, Cf_2S_3, body-centred cubic; [f] βUA_2; βUTe_2, orthorhombic, Immm.;
[g] γPu_2A_3, body-centred cubic, Th_3P_4; [h] Monoclinic, $CeSe_2$ (?); [i] fcc, face-centred cubic; [j] bcc, body-centred cubic;
[k] orth., orthorhombic; [l] Other crystallographically distinct phases are known.

phosphides (Chapter 5, p. 58), the common preparative route being by reaction of the finely divided metal with the chalcogen or hydrogen chalcogenide (sulphides and selenides only) at temperatures of 400–600°, an excess of chalcogen being required for the formation of compounds such as MA_3. These last are also frequently made by heating the dichalcogenides with the chalcogen at moderate temperatures, usually in a sealed tube.

It is also possible to precipitate some sulphides from fused salt solutions of the halide; an example is the precipitation of βUS_2 when hydrogen sulphide is passed through a solution of uranium tetrachloride in a fused NaCl/KCl eutectic. Another route to this compound is by heating uranium tetrachloride mixed with aluminium powder in hydrogen sulphide[28] at 350–550°:

$$3\ UCl_4\ +\ 4\ Al\ +\ 6\ H_2S\ \longrightarrow\ 3\ US_2(\beta)\ +\ 4\ AlCl_3\ +\ 6\ H_2$$

Such routes should also be practicable for the preparation of the chalcogenides of the transplutonium actinides, for which the metals are much less readily available than are thorium, uranium and plutonium. The sulphides of actinium[29] and of curium–californium[30] listed in Table 13 have only been prepared on the microgram scale, by reaction of the sesquioxide with a mixture of hydrogen sulphide and carbon disulphide at about 1100°. In the case of berkelium one might also expect the disulphide to be formed at a lower temperature, but this does not seem to have been investigated.

The chalcogenides MA_n, in which $n > 2$, are only known for thorium, uranium and neptunium; they decompose to dichalcogenides under vacuum above 900°. At higher temperatures (1300–1500°) all the known dichalcogenides decompose further. The magnetic moments of the uranium compounds of the types UA_3 and UA_2 are almost equal (~ 3.1 BM) and both compounds evidently contain U^{4+} ions, in contrast to UO_3 which is a compound of sexivalent uranium. Presumably, this difference is due either to the inability of uranium to form $(S=U=S)^{2+}$ cations analogous to the uranyl cation (possibly because of weaker $p\pi$–$d\pi$ interaction in the former) or to oxidation of the S^{2-} anion by the highly charged uranium cation. Some further evidence of uranium(IV) in US_3 is provided by the observation that the initial product of air oxidation of US_3 is uranium dioxide, and not the trioxide. Thorium, with a maximum of four valence electrons, can only be present as Th^{4+} ions in the corresponding telluride, $ThTe_3$; hence this compound, NpS_3 and the uranium trichalcogenides are best regarded as polychalcogenides.

The monochalcogenides, ThS, US and USe, behave as metallic-type conductors, whereas PuS exhibits semiconductor behaviour; in the case of the disulphides, ThS_2 is an insulator, its resistivity[31] being 10^{10} Ω cm, whereas βUS_2 has semiconductor properties, the resistivity[32] being only about 25 Ω cm. The resistivities of the other uranium dichalcogenides are of the same order as that of βUS_2, but data are lacking for the compounds of the other actinides.

Oxochalcogenides of the general types MOA (Th, U; A = S, Se, Te; Pa, Np, A = S) and M_2O_2A (U, A = Te; Pu, A = S, Se, Te) are known; they are made, for example, by heating together the actinide dioxide and dichalcogenide. The compounds of the type MOA all adopt the tetragonal PbFCl structure.

7. REFERENCES

1 E.H.P. Cordfunke, *The Chemistry of Uranium*, Elsevier, Amsterdam, 1969, Chapter 6, p. 98.
2 J.M. Cleveland, *The Chemistry of Plutonium*, Gordon and Breach, New York, 1970, Chapters 9 and 15.
3 L.E.J. Roberts, *Quart. Rev.*, 15 (1961) 442.
4 C. Keller, *Intern. Rev. Sci.*, 7 (1972) 47; Medical and Technical Publishing Co., and Butterworth, London.
5 R.M. Dell and N. Bridger, *Intern. Rev. Sci.*, 7 (1972) 211; Medical and Technical Publishing Co., and Butterworth, London.
6 Y. Akimoto, *J. Inorg. Nucl. Chem.*, 29 (1967) 2650.
7 W.H. Zachariasen, *Acta Cryst.*, 5 (1952) 19.
8 J. Williams and K.H. Westmacott, *Rev. Met.*, 53 (1956) 198.
9 J.L. Green and B.B. Cunningham, *Inorg. Nucl. Chem. Letters*, 3 (1967) 343.
10 M. Schlechter, J. Kooi, R. Billiau, R.A. Charlier and G.L. Dumont, *J. Nucl. Materials*, 15 (1965) 189.
11 L.E.J. Roberts and A.J. Walter, *Proc. Intern. Colloq. Physicochim. Protactinium*, Orsay, 1965, 51.

12 H. Blank and C. Ronchi, *Acta Cryst.*, A24 (1968) 657.
13 H.R. Hoekstra, S. Siegel, L.H. Fuchs and J.J. Katz, *J. Phys. Chem.*, 59 (1955) 136.
14 L.E.J. Roberts and A.J. Walter, *U.K. At. Energy Authority Rept.*, (1963) *A.E.R.E. R3624.*
15 J.-M. Leroy and G. Tridot, *Compt. Rend.*, C262 (1966) 114.
16 K.W. Bagnall and J.B. Laidler, *J. Chem. Soc.*, (1964) 2693.
17 D. Cohen and A.J. Walter, *J. Chem. Soc.*, (1964) 2697.
18 T. Stchouzkoy, H. Pezerat and R. Muxart, *Proc. Intern. Colloq. Physicochim. Protactinium*, Orsay, (1965) 61.
19 G. Gordon and H. Taube, *J. Inorg. Nucl. Chem.*, 16 (1961) 268.
20 J.J. Katz and D.M. Gruen, *J. Am. Chem. Soc.*, 71 (1949) 2106.
21 W.O. Milligan, M.L. Beasley, M.H. Lloyd and R.G. Haire, *Acta Cryst.*, B24 (1968) 979.
22 G. Bergstrom and G. Lundgren, *Acta Chem. Scand.*, 10 (1956) 673.
23 J.M. Cleveland, *Inorg. Nucl. Chem. Letters*, 6 (1970) 535.
24 V.I. Spitsyn, A.D. Gelman, N.N. Krot, M.P. Mefodiyeva, F.A. Zakharova, Yu. A. Komkov, V.P. Shilov and I.V. Smirnova, *J. Inorg. Nucl. Chem.*, 31 (1969) 2733.
25 J.-M. Leroy and G. Tridot, *Compt. Rend.*, C262 (1966) 1376.
26 S.K. Awasthi, L. Martinot, J. Fuger and G. Duyckaerts, *Inorg. Nucl. Chem. Letters*, 7 (1971) 145.
27 C. Keller and H. Seifert, *Inorg. Nucl. Chem. Letters*, 5 (1969) 51.
28 Ĉ. Yoshihara, M. Kanno and T. Mukaibo, *J. Nucl. Sci. Technol.*, 4 (1967) 578.
29 S. Fried, F. Hagemann and W.H. Zachariasen, *J. Am. Chem. Soc.*, 72 (1950) 771.
30 D. Cohen, S. Fried, S. Siegel and B. Tani, *Inorg. Nucl. Chem. Letters*, 4 (1968) 257.
31 F.K. McTaggart, *Aust. J. Chem.*, 11 (1958) 471.
32 P.K. Smith and L. Cathey, *J. Electrochem. Soc.*, 114 (1967) 973.
33 E.A. Moskal'kova and Yu. M. Tolmachev, *Soviet Radiochem.*, 12 (1970) 723.
34 C. Musikas, *Radiochem. Radioanalyt. Letters*, 4 (1970) 347.

Chapter 7

THE HALIDES AND PSEUDOHALIDES

1. INTRODUCTION

Chemical information on the actinide halides, and the complexes formed by them, make up the major part of the published literature concerned with the chemistry of the $5f$ elements; there have been several recent reviews of the halide chemistry[1-3], and there is one review which deals specifically with the coordination chemistry[4] of the actinide halides.

The known halides are listed in Table 14; from this it can be seen that the thermal stability towards reduction of the halides of the actinides in their higher oxidation states clearly decreases with increasing atomic number of the halogen. Thus UI_4 is rather easily converted to UI_3 by heat while neptunium and other actinide tetraiodides are unknown and must be even more unstable thermodynamically than the

TABLE 14

THE ACTINIDE HALIDES

MX_2	MX_3	MX_4	MX_5	MX_6
Ac, X = –	F, Cl, Br, I	–	–	–
Th, X = I	I	F, Cl, Br, I	–	–
Pa, X = –	I (?)	F, Cl, Br, I	F, Cl, Br, I	–
U, X = –	F, Cl, Br, I	F, Cl, Br, I	F, Cl, Br	F, Cl
Np, X = –	F, Cl, Br, I	F, Cl, Br	F	F
Pu, X = –	F, Cl, Br, I	F	–	F
Am, X = –	F, Cl, Br, I	F	–	–
Bk, X = –	F, Cl, Br, I	F	–	–
Cf, X = Br	F, Cl, Br, I	F	–	–
Es, X = –	Cl	–	–	–

uranium compound. There is, however, inferential evidence that NpI_4 may exist in a UI_4 matrix[5], for when neutron-irradiated uranium metal ($^{239}Np + {}^{238}U$) is converted to the tetraiodide in preparation for the reaction to form the phthalocyanine complex, neptunium tetraiodide may possibly be formed. Tetrachlorides and tetrabromides are known for the elements from thorium to neptunium, $NpBr_4$ being the least stable to heat of these compounds. Although plutonium tetrachloride and tetrabromide are unknown, it is possible to make anionic hexahalo-complexes in aqueous, or aqueous ethanolic, solution; these compounds can then be used in the preparation of other complexes of the tetrahalides. For example, reaction of caesium hexachloroplutonate(IV), Cs_2PuCl_6, with suitable oxygen-donor ligands in non-aqueous solvents yields complexes of $PuCl_4$. Oxygen-donor complexes of $PuBr_4$ are also known; they are easily prepared by oxidation of the corresponding complex of the tribromide in methyl cyanide with bromine.

The metals in the quinquevalent and sexivalent state exhibit a similar trend; all four pentahalides are known for protactinium, all but the pentaiodide are known for uranium, whereas only the penta-fluoride exists for neptunium; similarly, the only recorded hexahalides are the hexafluorides of uranium, neptunium and plutonium and the hexachloride of uranium.

This trend is a reflection of the increasing oxidising power of the higher oxidation states of the actinides as the atomic number of the metal increases, the only isolable compounds of the metals in these states being those in which the halide anions are the most highly electro-negative and hence least easily oxidised (p. 19).

In the following pages the halides and their complexes are grouped together by oxidation states, these being from +2 to +6. The discussion is deliberately brief and rather general; more detail is to be found in the reviews[1-4].

2. THE BIVALENT ACTINIDES

The only reported[6,7] actinide dihalides are ThI_2, obtained when the tetraiodide reacts with the stoicheiometric quantity of thorium metal at

600–800°, and $CfBr_2$. Two crystal modifications of ThI_2 are known; the compound disproportionates at high temperatures and reacts with water with the evolution of hydrogen. The diiodide is not a simple compound of Th^{2+}; it probably contains the Th^{4+} ion and is best represented[6] as $Th^{4+}(I^-)_2(2e^-)$. Its structure, which is consistent with this view, is made up of four two-dimensionally infinite layers, alternately trigonal prismatic and trigonal antiprismatic (Fig. 8). The actual coordination polyhedra of the thorium atoms are a trigonal antiprism of iodine atoms in the antiprismatic layers and a trigonal prism in the prismatic layers[155]. $CfBr_2$, an amber solid which is isomorphous with $EuBr_2$, is prepared[247] by reducing the tribromide in hydrogen at 650°. The reduction is slightly more difficult than that of $SmBr_3$. Nobelium, which appears to be normally bipositive in aqueous media (No^{2+}, f^{14}) should also form ionic dihalides. However, the Am^{2+} ion can be stabilized[8] in a CaF_2 matrix; it is easily identified by its ESR spectrum (Am^{2+}, f^7, ground state $^8S_{7/2}$; Am^{3+}, f^6, ground state $^7F_0$) and is produced from Am^{3+} in the CaF_2 either by its own self-irradiation or by the α-radiation from the shorter lived ^{244}Cm, which was also incorporated in the crystal. The Es^{2+} ion can be stabilized in the same way[9].

3. THE TERVALENT ACTINIDES

(a) Trifluorides

The known compounds for actinides in this oxidation state are listed in Table 15. Trifluorides of the actinides, up to berkelium trifluoride, which is dimorphic[10], have the LaF_3 type structure (hexagonal, $P6_3/mmc–D^4_{6h}$) in which the metal lies on a twofold axis with nine fluorine atoms arranged in a heptagonal bipyramid at almost equal distances[11], the overall symmetry being C_{2v} (Fig. 9). The second form of BkF_3, and the only known[10] form of CfF_3, have the orthorhombic YF_3 structure (space group $Pnma–D^{16}_{2h}$) in which[12] each metal atom has eight fluorine atoms arranged about it in the form of a trigonal prism with two of the fluorine atoms capping two of the three lateral faces of the prism. A ninth fluorine atom is situated somewhat further

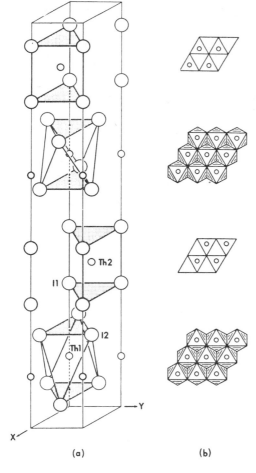

(a)   (b)

Fig. 8. View of the crystal structure of ThI_2 showing (a) the contents of one
unit cell and (b) [001] projections of four cells for each of the four layers in the
structure (L.J. Guggenberger and R.A. Jacobsen, *Inorg. Chem.*, 7 (1968) 2257).
Copyright (1968) by the American Chemical Society and reprinted by
permission of the copyright owner.

($\sim$ 30 pm) from the metal atom and caps the third lateral face of the
prism, the whole arrangement corresponding, very approximately, to a
tricapped trigonal prism. The corresponding structural change in the

TABLE 15

THE ACTINIDE TRIHALIDES[a]

AcF_3	$AcCl_3$	$AcBr_3$	AcI_3
white	white	white	
			ThI_3
			black
			PaI_3
			black
UF_3	UCl_3 (835)	UBr_3 (730)	UI_3 (766)
black	olive-green	red	black
NpF_3 (1425)	$NpCl_3$ (~800)	$NpBr_3$	NpI_3 (770)
purple	green	green	brown
PuF_3	$PuCl_3$ (760)	$PuBr_3$ (681)	PuI_3
violet-blue	emerald-green	green	bright-green
AmF_3 (1395)	$AmCl_3$	$AmBr_3$	AmI_3
pink	pink	white	yellow
CmF_3 (1406)	$CmCl_3$ (500)	$CmBr_3$ (400)	CmI_3
white	white	white	white
BkF_3	$BkCl_3$	$BkBr_3$	BkI_3
yellow-green	green	yellow-green	yellow
CfF_3	$CfCl_3$		CfI_3
	green		lemon-yellow

[a]Melting points (°C) in parentheses.

lanthanide trifluorides occurs between promethium (Pm^{3+}, 6-coordinate ionic radius 94.5 pm) and samarium (Sm^{3+}, ionic radius 92.1 pm), that is, at the same point as the change in the actinides, the ionic radius of Bk^{3+} being 93 pm.

These trifluorides are all insoluble in water and the addition of fluoride ion to a solution of the actinide(III) precipitates the trifluoride hydrate. Anhydrous higher actinide (Am–Cf) trifluorides can be prepared by heating the precipitated hydrate in hydrogen fluoride at 400°; the method is little used and the common preparative route is by heating the sesquioxides, M_2O_3 (Ac, Am, Cm, Cf), in hydrogen fluoride, or a mixture of hydrogen fluoride and hydrogen (Bk), at 600–700°.

Uranium, neptunium and plutonium trifluorides are more difficult to prepare because of the ease of oxidation to the tetrafluoride. UF_3 is best made by reducing the tetrafluoride with finely divided uranium metal[230] or with aluminium metal, the reduction occurring at 900° in

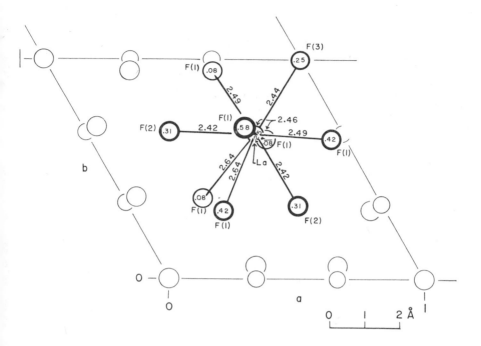

Fig. 9. The ninefold coordination of lanthanum in lanthanum trifluoride as viewed down the c axis (A. Zalkin, D.H. Templeton and T.E. Hopkins, *Inorg. Chem.*, 5 (1966) 1466). Copyright (1966) by the American Chemical Society and reprinted by permission of the copyright holder.

both cases. With aluminium, the monofluoride sublimes from the reaction mixture:

$$3\ UF_4\ +\ U \rightleftharpoons 4\ UF_3$$

$$UF_4\ +\ Al \rightarrow\ UF_3\ +\ AlF$$

Unfortunately, UF_3 disproportionates to the tetrafluoride and the metal at high temperature, the rate being detectible even at 800°. Reduction of UF_4 with magnesium at 560° also yields UF_3, but at higher temperatures the reaction goes a stage further and uranium metal is formed[13]. Such complications do not arise with NpF_3 and PuF_3, their preparation requiring quite mild reducing conditions. For example,

References p. 133

reaction of the appropriate dioxide with an equimolar mixture of hydrogen and hydrogen fluoride yields NpF_3 at $500°$ (ref. 14) and PuF_3 at $600°$ (ref. 15). The latter can also be prepared by dehydrating the hydrate, $PuF_3 \cdot 0.75 \, H_2O$, at $500°$, in hydrogen and hydrogen fluoride. This is a reaction mixture which seems to yield PuF_3 with almost any plutonium compound.

The trifluorides are high melting solids and even UF_3 is only slowly oxidised in air at room temperature; they are, however, converted to oxides at higher temperatures, UF_3 yielding U_3O_8 at $900°$ and PuF_3 forms PuO_2 at $300°$. UF_3 reacts only slowly with water at room temperature, but uranium(VI) is formed at $100°$; reaction of UF_3 with the other halogens is also slow even at moderately high temperatures, the product being a mixed halide[16] of the type UF_3X.

Fluorocomplexes of the type $NaMF_4$ (U, Pu, Am), obtained by melting together the stoicheiometric amounts of sodium fluoride and the actinide(III) fluoride, are known; in these compounds the actinide metal atom is surrounded by nine fluorine atoms, the arrangement being close to a symmetrically tricapped trigonal prism, as in the neodymium compound[17], $NaNdF_4$. The analogous potassium, rubidium and caesium tetrafluorouranates(III) are also known[18], prepared by heating together the stoicheiometric quantities of uranium tetrafluoride, uranium metal and alkali metal fluoride at $1000°$. Complex uranium(III) fluorides of composition $M_2^I UF_5$ and $M_3^I UF_6$ (M^I = K, Rb, Cs) have been made in the same way but, apart from the crystal symmetry of the purple-brown K_3UF_6, which is face-centred cubic, structural information is not available. However, it has been noted[18] that K_3UF_6 is isomorphous with α-K_3UF_7 and K_3UF_8, the cell parameters being almost identical for the three compounds.

(b) Trichlorides

There is a marked change in the arrangement of the chlorine atoms bonded to the actinide metal in the crystalline trichlorides as the ionic radius of the elements decreases across the series, paralleling the behaviour observed for the lanthanide trichlorides. For the largest actinide M^{3+} ions (Ac, U–Cf, and Es at 425°), the trichlorides have the hexagonal UCl_3 (or $Y(OH)_3$) type structure (space group $C6_3m$-C_{6h}^2) in which each actinide atom is surrounded by nine chlorine atoms arranged in the form of a symmetrically tricapped trigonal prism. The nine M–Cl distances are almost equal in UCl_3, but distances for the six atoms in the prism diverge noticeably from the values for the three capping the lateral faces of the prism as the atomic number increases. Consequently values for the M^{3+} ionic radii obtained by averaging the nine M–Cl distances in the trichlorides are somewhat suspect[19].

Californium trichloride is, however, dimorphic, the second crystal form having the orthorhombic (space group $Ccmm$-D_{2h}^{17}) $PuBr_3$ type structure, in which the metal is bonded to eight halogen atoms arranged in the form of a trigonal prism with two capped lateral faces. There should also be a low temperature form of $EsCl_3$ with this type of structure, but this form has not yet been reported. The higher actinide trichlorides may well adopt the six coordinate $AlCl_3$ type of structure observed for the trichlorides of the lanthanides of smallest radius, but no data for these are yet available. The corresponding structure changes in the lanthanides occur between gadolinium (UCl_3 type) and terbium ($PuBr_3$ type), the 6-coordinate ionic radii being 90 and 88.8 pm respectively, as compared with 91.5 pm for Cf^{3+}, and between terbium and dysprosium (ionic radius 87.7 pm).

The actinide trichlorides are hygroscopic, forming hexahydrates (Pu, Am, Bk) in which $[MCl_2(H_2O)_6]^+$ cations are present in the lattice[231], and all are soluble in water. The anhydrous compounds melt at temperatures between 760° ($PuCl_3$) and 835° (UCl_3); the reported[20] melting point of $CmCl_3$, 500°, appears to be rather low compared with values for the lanthanide trichlorides, but no data are available for $AmCl_3$ or $CfCl_3$ for comparison. All the known trichlorides can be sublimed in a vacuum at high temperatures and vapour pressure data have been summarised by Brown[2].

The trichlorides of the earlier actinides, up to americium, have been prepared by very many different routes; some of the more useful methods are by the reaction of the metal hydride with hydrogen chloride at 250° (UCl_3) or 400° ($PuCl_3$), a method which should be applicable also to $NpCl_3$ and the transplutonium actinides when the

References p. 133

metals become more readily available, and by the reaction of carbon tetrachloride with the hydroxide ($Ac(OH)_3$ up to $960°$) or dioxide (AmO_2 at $800-900°$). The reaction of the dioxide with hydrogen chloride (AmO_2 at $500°$; PuO_2 at $200-1000°$) yields $AmCl_3$ or $PuCl_3$ and $PuOCl$. Reduction of the tetrachloride with hydrogen (UCl_4, under pressure, at $550-650°$; $NpCl_4$ at $450°$) or with metals such as zinc (UCl_4 at $450-480°$) is also practicable. For large-scale production of $PuCl_3$, the best routes seem to be by the reaction of plutonium(III) or plutonium(IV) oxalate with hydrogen chloride at $140-500°$, or plutonium(III) carbonate with carbonyl chloride at $500-550°$. Another useful method consists in evaporating a hydrochloric acid solution of the tripositive actinide (applicable to Am, Cm), in the presence of ammonium chloride, to dryness and then subliming the trichloride from the residue at high temperature in a vacuum; the reaction of actinium hydroxide with ammonium chloride at $250°$ also yields $AcCl_3$. The reaction of ammonium chloride with the hydroxide or with the hydrated chloride is essentially the same process as dehydrating the latter in hydrogen chloride, another possible mode of preparation.

For curium and the higher actinides, of which only small (generally sub-microgram) amounts of the elements are normally available, the common preparative route for the trichlorides is by allowing the sesquioxide, M_2O_3, to react with hydrogen chloride at $450-500°$. A fuller list of methods available for the preparation of the trichlorides can be found in the reviews[1,2].

The chemical behaviour of the trichlorides has not been investigated extensively. Uranium trichloride is known to disproportionate[21] above $840°$ and it reacts, like UF_3, with halogens of higher atomic number than chlorine to form mixed halides of the type UCl_3X; neptunium trichloride should certainly react with bromine to form an analogous compound. UCl_3 also reacts with gaseous ammonia at $450-500°$ to yield amidocompounds, $U(NH_2)Cl_2$ and $U(NH_2)_2Cl$, which decompose[22] above $800°$ to form the nitride, $UN_{1.73-1.75}$, and the imidochloride, $U(NH)Cl$, which is isostructural[23] with the oxochloride. Neptunium trichloride does not seem to undergo ammonolysis, for the tetrachloride is reduced[24] to the trichloride in gaseous ammonia at $350-400°$; it may be that ammonolysis occurs only at a higher temperature, however.

Very few adducts have been reported for uranium trichloride; the ammines are of composition $UCl_3 \cdot 6.6\text{-}7.4\ NH_3$, and it has been shown[241] that only three of the ammonia molecules are coordinated to the uranium atom. A methyl cyanide adduct, $UCl_3 \cdot CH_3CN$, is obtained by treating the trichloride with the oxygen-free ligand at 80°; the compound is isostructural with $CeCl_3 \cdot CH_3CN$. The reaction leading to the formation of this adduct is apparently very slow[242]. Phenazone (Pz, $C_{11}H_{14}N_2O$), a cyclic compound in which the ring consists of three carbon and two nitrogen atoms, acts as an oxygen donor ligand by way of a ring carbonyl group and yields the purple complex $U(Pz)_6Cl_3$ when an ethanolic solution of the ligand is added to a solution of $M^IUCl_4 \cdot 5\ H_2O$ (M^I = K, Rb) in dry ethanol[246].

Few anionic complexes of the trichlorides have been reported; chlorocomplexes of the type $Cs_2NaM^{III}Cl_6$ (M = Am[25], Bk[26]), obtained from aqueous hydrochloric acid solution when sodium and caesium ions are present, are known and both compounds are of face-centred cubic symmetry (space group $Fm3m\text{-}O_h^5$). The phosphonium salts $[(C_6H_5)_3PH]_3M^{III}Cl_6$ (M = Pu, Am) are formed in nitromethane or ethanol[27]. $CsAmCl_4 \cdot {\sim}4\ H_2O$ and Cs_3AmCl_6 have also been prepared from concentrated and from ethanolic hydrochloric acid respectively[25, 28]; with the tetraethylammonium ion, $(NEt_4)_3AmCl_6$ and $(NEt_4)_2LiAmCl_6$ can be isolated[28]. The only other chlorocomplexes obtained from aqueous solution are, rather surprisingly in view of the ease of oxidation, the purple hydrated salts $M^IUCl_4 \cdot 5\ H_2O$ (M^I = NH_4, K, Rb), isolated[29] from cold, concentrated hydrochloric solutions of the uranium(III) sulphatocomplexes (p. 163). The report[30] of the precipitation of a hexachloroplutonate(III), $Cs_3PuCl_6 \cdot 2\ H_2O$, has not been confirmed[25, 26], but this parallels the behaviour of uranium(III), for which a caesium salt of the chlorocomplex could not be obtained; however, it is probable that neptunium(III) and plutonium(III) analogues of the uranium(III) chlorocomplex mentioned above could be prepared under the correct conditions.

There is evidence for the formation of chlorocomplex species, such as $PuCl_5^{2-}$, $PuCl_6^{3-}$, $PuCl_9^{6-}$ and $Pu_2Cl_7^-$ in fused salt melts, but it is very likely that many of these can exist only in such media.

(c) *Tribromides*

The structures of the tribromides, like those of the trichlorides, change from the UCl_3 type (Ac, U, $\alpha NpBr_3$) to the $PuBr_3$ type ($\beta NpBr_3$, Pu–Bk) with decreasing M^{3+} radius. $CfBr_3$, however, exists in a third structural type, which is not yet positively identified[31]. The existence of $\beta NpBr_3$ has been questioned, for the phase was not observed in a recent study[82], but the exact details of the method used in the original preparation are lacking. The change in the structure of the tribromides occurs between praseodymium and neodymium in the lanthanides (6-coordinate radii 96.8 and 95.5 pm respectively) so that the change in the actinide series at neptunium (radius 98.6 pm) is by no means impossible if one compares the difference in change-over points in the corresponding trichlorides. The highest actinides should also follow the lanthanides and adopt the octahedral, 6-coordinate arrangement of the $FeCl_3$ structure, probably, on radius considerations, at or near californium.

Like the trichlorides, the tribromides are hygroscopic and soluble in water; the hexahydrated tribromides, obtained by exposing the anhydrous compound to moist air, are isostructural with their chloride analogues and lose water (U, Np, Pu, Am) in a vacuum, yielding the anhydrous compound[32]. This makes a convenient preparative method for anhydrous $AmBr_3$, for the hydrate is easily obtained by evaporating to dryness a solution of americium(III) in hydrobromic acid.

The common preparative routes are by the reaction of hydrogen bromide with the hydride (U at $300°$, Pu at $600°$), hydroxide (Pu(IV) at $600°$), oxalate hexahydrate (Pu(IV) at $500°$) or oxide (Bk, Cf at $800°$). Alternatives are reaction of the oxide with aluminium bromide (Ac_2O_3 at $750°$; NpO_2, in the presence of aluminium metal, at $300–400°$; AmO_2 at $500°$) or of the trichloride with ammonium bromide at $400–450°$ (Am, Cm). The known tribromides melt[33] between $400°$ ($CmBr_3$) and $730°$ (UBr_3), the melting point of the curium compound again being low compared with the lanthanides or the other actinides. The uranium compound disproportionates above $900°$, and appears to form a hexammine with dry gaseous ammonia. Triphenylphosphonium salts of the hexabromocomplexes can be isolated[27] from ethanolic hydrogen bromide solutions of plutonium(III) and americium(III), but apart from these compounds and the ammine mentioned above no complexes are known.

(d) Triiodides

Although thorium forms a black compound of analytical composition ThI_3, prepared[7] from ThI_4 and the stoicheiometric quantity of thorium metal at 600-700°, this is best regarded as a thorium(IV) compound, Th^{4+} $(I^-)_3(e^-)$; it reacts vigorously with water with the evolution of hydrogen. Protactinium triiodide is reported[34] to be formed when milligram amounts of the pentaiodide are heated under vacuum with continuous pumping; the identification rests on the X-ray powder data, which indicate that the product is isostructural with $CeI_3(PuBr_3$ type). The nature of this product is not fully established and further work on this system is still required.

The triiodides of the earlier actinides (Pa, above, and U–αAm) adopt the 8-coordinate arrangement of the $PuBr_3$ type structure (p. 85), whereas the later members of the series (βAm–Cf) crystallise with the hexagonal BiI_3 lattice (space group $R\bar{3}-C_{3i}^2$) in which the metal atom is octahedrally surrounded by six iodine atoms. The analogous structural change in the lanthanides occurs between neodymium and samarium, for which the 6-coordinate ionic radii are 95.5 and 92.1 pm respectively (the structure of PmI_3 appears to be unknown), compared with 96.2 pm for the radius of Am^{3+}, which appears to form triiodides of both structure types.

The simplest ways of preparing the triiodides are by the reaction of the metal with iodine at about 500° (U^{35}, Np^{36}), with mercuric iodide at 500° (Pu^{37}) or hydrogen iodide at 450° (Pu^{38}). Many of the actinide metals, however, are not commonly available, so that iodination of an oxide is often the only possible route. Successful reactions include that of the oxide with aluminium iodide (Ac, Np, Am) or ammonium iodide (Ac) above 500°, and with hydrogen iodide at about 650° (Bk^{39}, Cf^{31}), but the use of halogen exchange reactions for the preparation of these compounds, for example by heating the trichlorides with ammonium iodide (Am^{37}, Cm^{40}) is also attractive.

The triiodides, like the chlorides and bromides, are soluble in water, but complexes derived from the triiodides appear to be unknown; the melting points are about 770° (U–Pu), comparable with those of the lanthanide triiodides, and they can be sublimed in a vacuum.

(e) Oxohalides, MOX

Oxohalides of composition MOX, where X = F (Ac, Pu, Cf), Cl (Ac, U, Pu–Es), Br (Ac, Pu, Cm–Cf) and I (Pu, Bk, Cf) have been recorded. AcOF and CfOF have the cubic, CaF_2 type structure (space group $Fm3m-O_h^5$) in which the oxygen and fluorine atoms are arranged randomly in the lattice, whereas the remaining oxohalides have the tetragonal PbFCl structure (space group $P4/nmm-D_{4h}^7$), a layer structure in which the metal atoms lie between coplanar sheets of halogen and oxygen atoms, each metal atom being surrounded by 4 oxygen and 4 halogen atoms. These compounds are prepared by controlled (or accidental!) hydrolysis of the trihalides and little is known about their chemical properties.

4. THE QUADRIVALENT ACTINIDES

The tetrahalides (Table 16) are the best known of all actinide compounds, particularly with regard to the complexes formed by the chlorides with oxygen donor ligands, which have frequently been described in some detail in the reviews[1,3,4].

(a) Tetrafluorides

Tetrafluorides are known for the first nine actinides (Th–Cf); all are of monoclinic symmetry (space group $C2/c-C_{2h}^6$), isomorphous with the known lanthanide (Ce, Pr, Tb) tetrafluorides. The metal is 8-coordinate in this structure, each metal atom being surrounded by eight fluorine atoms in a somewhat distorted antiprismatic arrangement[41].

The most convenient preparative route is by the reaction of the dioxides with hydrogen fluoride, either alone (Th, U), mixed with hydrogen (Pa) or mixed with oxygen (Np, Pu), at 500–600°. The higher

TABLE 16

THE ACTINIDE TETRAHALIDES[a]

ThF_4 (1110)	$ThCl_4$ (770)	$ThBr_4$ (679)	ThI_4 (566)
white	white	white	yellow
PaF_4	$PaCl_4$	$PaBr_4$	PaI_4
brown	greenish-yellow	red	dark green
UF_4 (1036)	UCl_4 (590)	UBr_4 (519)	UI_4 (506[b])
green	green	brown	black
NpF_4	$NpCl_4$ (538)	$NpBr_4$ (464)	
green	red-brown	red-brown	
PuF_4 (1027)			
brown			
AmF_4			
tan			
CmF_4			
greenish-tan			
BkF_4			
CfF_4			

[a] Melting points (°C) in parentheses.
[b] In an atmosphere of iodine.

actinides (Am–Cf) require more forcing conditions, the normal preparative route being reaction of the trifluoride with fluorine at about 500°; it should be noted that, in the case of curium, formation of the tetrafluoride only occurs with the longer-lived isotopes, such as ^{244}Cm; radiation damage appears to prevent its formation when the short-lived ^{242}Cm is used. Possible alternative routes include reaction of the dioxides (U at 500°, Pu at 600°) with sulphur tetrafluoride and dehydration of the precipitated tetrafluoride hydrates (Th, Pu) in hydrogen fluoride. Summaries of these and other published preparative methods are available[1-4].

   These compounds (Th–Pu) are insoluble in water (presumably the remaining tetrafluorides will also be insoluble) and addition of fluoride ion to an aqueous solution of an actinide(IV) precipitates the hydrated compound $MF_4 \cdot 2.5 H_2O$ (Th, U, Pu); protactinium(IV) also yields a precipitate under these conditions but its composition is uncertain. Neptunium(IV) should behave in the same way, but does not seem to have been investigated. The americium(IV) and curium(IV) tetrafluoride hydrates cannot be prepared by precipitation because the quadripositive ions of these elements are unstable with respect to disproportionation (Am) and reduction (Cm) in water.

The precipitation of the hydrated tetrafluorides (Pa, U, Pu) from sulphate solution appears to occur by way of an intermediate MF_2^{2+} species and the compound $PaF_2(SO_4) \cdot 2\,H_2O$ has been isolated[42]. The somewhat similar oxalatofluoride, $UF_2(C_2O_4) \cdot 1.5\,H_2O$, has been prepared[43] by heating the hydrated tetrafluoride with saturated oxalic acid solution at $100°$. Analogous compounds of neptunium and plutonium have not been reported as yet, but should be capable of existence.

(b) Fluorocomplexes

The stoicheiometries of the principal types of complex are $M^IM^{IV}F_5$, $M_2^IM^{IV}F_6$ (and $M^{II}M^{IV}F_6$), $M_3^IM^{IV}F_7$, $M_4^IM^{IV}F_8$ and $M_7^IM_6^{IV}F_{31}$. Examples of most of these types of complex have been recorded for all the elements from thorium to berkelium, but not all possible combinations of alkali metal (including the ammonium ion) and actinide metal can exist.

In the structure of $LiUF_5$ (tetragonal, space group $I4_1/a$-C_{4h}^6), with which the other lithium compounds (Th–Cm) are isomorphous, the uranium atom is surrounded by nine fluorine atoms in a tricapped trigonal prismatic array, resembling the UCl_3 type structure in geometry. It appears[44] that compounds of this composition only exist when the ratio of the radii M^+/M^{4+} is less than 0.99 or greater than 1.59, while for radius ratios between 0.99 and 1.59 the compounds $M_7^IM_6^{IV}F_{31}$ are formed. Both of the above types of fluorocomplex exist at radius ratios between 1.59 and 1.68.

The compounds $M_7^IM_6^{IV}F_{31}$ are of rhombohedral symmetry (space group $R\bar{3}$-C_{3i}^2), isostructural with $Na_7Zr_6F_{31}$, in which the six quadripositive metal atoms are at the apices of an octahedron; each of the metal atoms is surrounded by eight fluorine atoms in an antiprismatic arrangement, the six antiprisms sharing corners to form the octahedral arrangement in which the remaining fluorine atom is situated[45].

In some of the hexafluorocomplexes the geometry of the environment of the actinide metal atom approximates to a tricapped trigonal prism (β_2-Na$_2$ThF$_6$, β_1-K$_2$UF$_6$), and in others, such as the alkaline earth metal compounds, M^{II}M^{IV}F$_6$ (Th–Pu), the 9-coordinate arrangement is a disordered version of the LaF$_3$ structure; in a third group (e.g. αK$_2$UF$_6$) the actinide metal is 8-coordinate in a deformed cubic arrangement, while in Rb$_2$UF$_6$ the structure consists of infinite chains of UF$_8$ dodecahedra[245]. However, in the compounds M$_3^I$M^{IV}F$_7$ (e.g. Na$_3$UF$_7$) the coordination geometry is a simple pentagonal bipyramid[46] and in M$_4^I$M^{IV}F$_8$ the actinide metal is 8-coordinate, the arrangement being a distorted tetragonal prism[47].

Many of these fluorocomplexes can be prepared by precipitation from aqueous solutions of the actinide(IV), examples of complexes produced in this way being (NH$_4$)$_4$M^{IV}F$_8$ (U, Pu, Am), some of the M$_7^I$M$_6^{IV}$F$_{31}$ (e.g. (NH$_4$)$_7$U$_6$F$_{31}$) and most of the compounds M$_2^I$M^{IV}F$_6$. A more generally applicable route is by heating together, in stoicheiometric proportions, the actinide tetrafluoride and alkali metal fluoride; alternatively, the actinide metal dioxide and alkali (or alkaline earth) metal carbonate are heated together in hydrogen fluoride, with oxygen present in the case of the plutonium compounds. A third, somewhat similar, method consists in evaporating to dryness an aqueous acid solution of the actinide(IV) and alkali metal (in the stoicheiometric proportions for the compound required), followed by heating the residue in hydrogen fluoride or fluorine (Am, Cm).

A much fuller account of these, and other, actinide(IV) fluorocomplexes has been given in a recent book[2], which included much more preparative detail than is given above; there is also an even more up-to-date account of the halides and halo-complexes formed by the trans-plutonium elements[48].

(c) Tetrachlorides

Tetrachlorides are known only for the first four elements (Th–Np), but there is some evidence for the existence of PuCl$_4$ in the gaseous phase in equilibrium with PuCl$_3$ and chlorine, and its electronic spectrum has been measured[49] in the gas phase at ~1000°. The four tetrachlorides are of tetragonal symmetry (space group $I4_1/amd\text{-}'D_{4h}^{19}$), eight chlorine atoms being arranged about the metal atom in a dodecahedron[50] (Fig. 10).

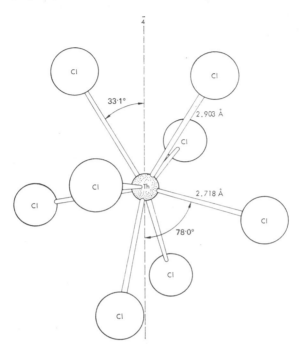

Fig. 10. The coordination arrangement in $ThCl_4$; the dodecahedron has $\overline{4}2m$ symmetry (K. Mucker, G.S. Smith, Q. Johnson and R.E. Elson, *Acta Cryst.*, B25 (1969) 2362). Reproduced by permission of the International Union of Crystallography.

A generally applicable route for small-scale preparations of the tetrachlorides is the reaction of the dioxide (Th, Pa, U, Np) with carbon tetrachloride vapour at about $500°$, but the dioxides used must be in a reasonably reactive form (such as those made by low temperature (450-500°) reactions) for the preparation to succeed. The method has been used in the production of large scale (kg) batches[51] of UCl_4. Because of the inertness of PaO_2, a more convenient route to $PaCl_4$ is by reduction of the pentachloride with aluminium metal[52]. The thorium compound is also made from the elements at $800°$ and by the reaction of thorium hydride with hydrogen chloride at $250-350°$; hydrated thorium tetrachloride, obtained from aqueous hydrochloric acid solution, can also be dehydrated by heating it with thionyl chloride. UCl_4 is commonly prepared by liquid phase chlorination of UO_3 or U_3O_8 with hexachloropropene[53] ($Cl_2C = CClCCl_3$), which boils at $210°$; the analogous reaction with Np_2O_5 or $NpO_3 \cdot H_2O$ yields $NpCl_4$ as well as some[54] $NpCl_3$. Uranium metal reacts with chlorine to yield a mixture of UCl_4 and higher chlorides, but since the latter are decomposed to UCl_4 by heat, the method is quite practicable for the preparation of the tetrachloride.

The tetrachlorides are appreciably more volatile than the trichlorides, subliming in a vacuum at about $500°$. They are hygroscopic, soluble in polar solvents, in which complexes are formed, and insoluble in non-polar solvents. They react with hydrogen fluoride to form tetrafluorides (Th, U) and with fluorine thorium and uranium tetrachlorides form the tetra- and hexafluorides respectively; uranium tetrachloride yields UO_2Cl_2 when heated in oxygen at $300°$, a reaction which should occur also for the neptunium compound at moderate temperatures, but at higher temperatures ($\sim 750°$) oxides are formed (U, Np). Complex species such as MCl_3^+ and MCl_2^{2+} appear to be present in aqueous hydrochloric acid solutions of thorium(IV) and plutonium(IV), and in aqueous ethanolic solutions of uranium tetrachloride.

(d) Chlorocomplexes

Hexachlorocomplexes, $M_2^I M^{IV} Cl_6$ (Th-Pu) are the principal chlorocomplex species known for the actinides(IV); the caesium and tetramethylammonium salts are of trigonal (space group $C3m-D_{3d}^3$) and facecentred cubic symmetry (space group $Fm3m-O_h^5$) respectively; in all of these compounds the MCl_6^{2-} ion is octahedral. Provided that the univalent cation is sufficiently large (e.g. Cs^+, R_4N^+, R_4As^+), the hexachlorocomplexes (Pa-Pu) are precipitated from concentrated hydrochloric acid solutions of the actinide(IV); these compounds are not hygroscopic. Where the cation is small (e.g. Li^+-K^+, $Ca^{2+}-Ba^{2+}$), the uranium compounds can be made by passing UCl_4 vapour over the appropriate solid

chloride at red heat[55]. The anhydrous thorium(IV) complexes are difficult to obtain pure from aqueous media, and are usually prepared by
heating together the stoicheiometric quantities of thorium tetrachloride
and the alkali metal chloride[56](Li–Cs). The analogous plutonium compounds (Na–Cs) are obtained in a similar manner, by heating a melt of
plutonium trichloride and the alkali metal chloride in chlorine[57].

The tetraethylammonium salts can also be made from the tetrachlorides (Th, U) and the ammonium chloride in thionyl chloride or
methyl cyanide; the latter solvent has been used for the preparation of
the protactinium(IV) complex.

Hydrated pentachlorothorates(IV), $M^IThCl_5 \cdot nH_2O$ (Li–K, NH$_4$),
can apparently be isolated from aqueous solution[56], and there is an
early report[58] of the formation of the complex KTh_2Cl_9, also obtained
from aqueous media; octachlorothorates(IV), $M_4^IThCl_8$ (Rb, Cs), are said
to be formed when the stoicheiometric quantities of thorium tetrachloride and the alkali metal chloride are fused together; it was reported
that smaller cations (Li–K) did not yield analogous compounds[56]. There
are no structural data available for any of these compounds and these
systems are worth reinvestigation.

(e) Oxygen-donor complexes

A very large number of oxygen-donor complexes are known and a
selection of these is given in Table 17. Where the ligand is bulky (e.g.
$(C_6H_5)_3PO$, $(CH_3)_3C \cdot CON(CH_3)_2$, $[(CH_3)_2N]_2CO$) the stoicheiometry
is usually $MCl_4 \cdot 2$ L where L represents the ligand. Magnetic and UV/
visible spectral data indicate that these most probably have a *trans*
octahedral configuration[69]. With smaller ligands, a considerable variation in stoicheiometry is observed, as can be seen from Table 17, and
since there are no X-ray structural data for these compounds, the
geometry of the complexes is a matter of conjecture.

Some trends are, however, apparent across the group in the cases
of the amide, sulphoxide and urea complexes; considering the
N,N-dimethylacetamide (DMA) complexes, the thorium compound seems to
be a monomer and the metal is probably 8-coordinate. The *tris*-DMA
protactinium complex degrades fairly readily to 2 $PaCl_4 \cdot 5$ DMA, the

TABLE 17

SOME OXYGEN DONOR COMPLEXES OF THE ACTINIDE TETRACHLORIDES

Amides, RCONR'$_2$	*Complex*	
$CH_3CON(CH_3)_2$	$MCl_4 \cdot 4$ L	M = Th[65]
(DMA)	$MCl_4 \cdot 3$ L	M = Pa[66]
	$2MCl_4 \cdot 5$ L	M = Pa[66], U-Pu[67]
Phosphine oxides, R$_3$PO		
$(C_6H_5)_3PO$ (TPPO)	$MCl_4 \cdot 2$ L	M = Th[59], U[60], Np[61], Pu[64]
$[(CH_3)_2N]_3PO$ (HMPA)	$MCl_4 \cdot 2$ L	M = Th[62], Pa[87], U[62], Np[61], Pu[64]
$[((CH_3)_2N)_2PO]_2O$ (OMPA)	$2MCl_4 \cdot 3$ L	M = U[63], Np[64], Pu[64]
	$MCl_4 \cdot 2$ L	M = Th[63]
Sulphoxides, R$_2$SO		
$(CH_3)_2SO$	$MCl_4 \cdot 7$ L[a]	M = U-Pu[68]
(DMSO)	$MCl_4 \cdot 5$ L	M = Th-Np[68]
	$MCl_4 \cdot 3$ L	M = Th, U-Pu[68]
Ureas		
$((CH_3)_2N)_2CO$	$MCl_4 \cdot 3$ L[b]	M = Th[69]
(TMU)	$MCl_4 \cdot 2$ L[b]	M = U[69]
$[(CH_3)NH)]_2CO$	$MCl_4 \cdot 6$ L	M = Th[69]
(DMU)	$MCl_4 \cdot 4$ L	M = U[69]

[a] Free ligand present in the lattice.
[b] Tetraethylurea and symmetrical N,N'-dimethyl, N,N'-diphenylurea form similar complexes[69].

stoicheiometry exhibited in the complexes formed by the remaining members of the group, but the structure of this has not been established. Thermal decomposition and chloride replacement studies by the present author[67,70] suggest that the compounds are chlorine-bridged dimers with one molecule of ligand in the bridge, but the UV/visible spectrum of a solution of the complex in methyl cyanide, in which the complex is virtually a non-electrolyte, is consistent with the presence of two differently coordinated uranium species, one octahedral and one of high coordination number and low symmetry.

The changes in the sulphoxide complexes are most noticeable in the *pentakis* species; the thorium and protactinium complexes are difficult to degrade to the *tris* complexes, whereas the uranium complex degrades easily, the neptunium complex is extremely unstable with respect to such degradation and the *pentakis* plutonium complex could not be obtained at all, indicating that the Pu^{4+} ion is too small to accommodate the five sulphoxide ligands and four chloride ions. The overall behaviour is consistent with the decrease in M^{4+} radius from thorium to plutonium. In the case of the urea complexes the work would have to be extended to the neptunium and plutonium complexes in order to complete the trend, but there is clearly a difference between the complexes formed by thorium and by uranium tetrachlorides.

Many other complexes of the actinide tetrachlorides have been recorded and information on these can be obtained from the reviews[1, 3].

The known complexes are easily prepared from the actinide tetrachloride by treatment with the ligand in a non-aqueous solvent; because plutonium tetrachloride does not exist in the solid state, complexes derived from it are prepared by treating the hexachlorocomplex, Cs_2PuCl_6, with the ligand[67, 68], and this method can be used for the uranium and neptunium tetrachloride complexes, so eliminating the necessity of working with the hygroscopic tetrachlorides.

(f) Sulphur-donor complexes

Complexes of the tetrachlorides with sulphur-donor ligands are very rare; the uranium tetrachloride complex with 1,2-dimethylthioethane[71] ($UCl_4 \cdot 2$ L) is probably the only example of this class of compound.

(g) Nitrogen-donor complexes

Complexes with ammonia, amines or nitriles are almost entirely restricted to thorium and uranium tetrachlorides. $ThCl_4 \cdot 6$ NH_3 is said[72] to be formed when dry ammonia gas is passed over solid thorium tetrachloride, whereas $ThCl_4 \cdot 8$ NH_3 is precipitated from ethereal solutions of the tetrachloride by gaseous ammonia; under similar conditions aliphatic amines precipitate complexes of the type[72] $ThCl_4 \cdot 4$ L. Uranium tetrachloride forms the ammines $UCl_4 \cdot 4$ NH_3, stable[73] to 45°, and[74] $UCl_4 \cdot 2$ NH_3; complexes with hydrazine, $UCl_4 \cdot 6$ N_2H_4, and primary amines[74], $UCl_4 \cdot L$ and $UCl_4 \cdot 2$ L, as well as with ethylenediamine, $UCl_4 \cdot 4$ *en*, piperidine, $UCl_4 \cdot 4$ L, and pyridine, $UCl_4 \cdot 2$ *py*, are also known[75]. The methyl cyanide complexes $MCl_4 \cdot 4$ L ($Th^{76, 77}$, U^{76}, Np^{61}) are made from the components, but the composition of the thorium tetrachloride complex with benzyl cyanide, reported as[78] $ThCl_4 \cdot 2$ L or[77] $ThCl_4 \cdot 4$ L, is somewhat uncertain. Structural data are not available for any of these compounds.

(h) Phosphorus-donor complexes

Complexes of the actinide tetrachlorides with phosphines are unknown; none of the reported preparations has been substantiated.

(i) Tetrabromides

Thorium tetrabromide is dimorphic, the high temperature β-form being isostructural with[79] UCl_4 and[52] $PaBr_4$. The detailed structures of the low temperature, orthorhombic[80] α-form and of[81] UBr_4, which is isostructural with[82] $NpBr_4$ (monoclinic), have not been reported.

The tetrabromides (Th, U, Np) are best made from the elements, although the reaction[83] of UO_3 with carbon tetrabromide at 165° appears to be a useful alternative for making UBr_4. Protactinium tetrabromide is best made by reduction of the pentabromide with aluminium metal[52] at 400–450°. The melting points of these compounds decrease fairly regularly, from 679° ($ThBr_4$) to 464° ($NpBr_4$); they are all quite volatile, subliming in a vacuum at 550° (Th–U), but neptunium tetrabromide decomposes to a considerable extent when sublimation is attempted in this way. All four compounds are hygroscopic, soluble in polar solvents in which complexes are formed, and insoluble in non-polar solvents. Hydrates, such as $ThBr_4 \cdot 7$ H_2O and $UBr_4 \cdot 8$ H_2O, obtained from hydrobromic acid solutions of the actinide(IV), are reported in the early literature, but little is known about them.

(j) Bromocomplexes

Only the hexabromocomplexes, in which the MBr_6^{2-} anion is octahedral, are known. The tetraethylammonium salts (Th–U) are prepared from methyl cyanide solutions of the component bromides[84,85], or (U, Np, Pu) from ethanolic hydrobromic acid solutions[86] of the actinide(IV).

(k) Oxygen-donor complexes

Some of the known oxygen-donor complexes of the tetrabromides are given in Table 18; as with the chloride complexes, structural data are generally not available, but from the spectrum of $UBr_4 \cdot 2$ TPPO it appears that the molecular geometry in this compound is octahedral.

TABLE 18

SOME OXYGEN-DONOR COMPLEXES OF THE ACTINIDE TETRABROMIDES

Amides		
$CH_3CON(CH_3)_2$	$MBr_4 \cdot 5$ L	Th-U[66]
(DMA)	$2MBr_4 \cdot 5$ L	Pa, U[66]
Phosphine oxides		
$(C_6H_5)_3PO$	$MBr_4 \cdot 2$ L	U[60], Np, Pu[61]
(TPPO)		
$[(CH_3)_2N]_3PO$	$MBr_4 \cdot 2$ L	Th[62], Pa[87], U[62], Np, Pu[61]
(HMPA)	$MBr_4 \cdot 3$ L	Th[62]
Sulphoxides		
$(CH_3)_2SO$	$MBr_4 \cdot 6$ L	Th, U[62]
(DMSO)		
Ureas		
$[(CH_3)_2N]_2CO^a$	$MBr_4 \cdot 3$ L	Th[69]
(TMU)	$MBr_4 \cdot 2$ L	U[69]

aTetraethyl and N,N'-dimethyl,N,N'-diphenyl ureas form similar uranium complexes[69].

The compounds (Th–Np) are prepared in the same way as the analogous tetrachlorides, using non-aqueous solvents. Conductivity data[62] for $ThBr_4 \cdot 6$ DMSO indicate that this compound behaves as a 1 : 1 electrolyte in nitromethane, the cation, presumably $[ThBr_3 \cdot 6 \text{ DMSO}]^+$, perhaps involving 9-coordinate thorium. $ThBr_4 \cdot 3$ HMPA is also an interesting compound, for it is the only tetrahalide–HMPA complex to be soluble in benzene, in which a molecular weight determination indicates that it is monomeric, but it also behaves as a 1 : 1 electrolyte in nitromethane[62]. It would certainly be of interest to know the coordination geometry. The compound decomposes to $ThBr_4 \cdot 2 HMPA$ when heated in a vacuum, and this last, like the analogous $MCl_4 \cdot 2 HMPA$, can be sublimed at about 250° in a non-static vacuum[62]. The plutonium tetrabromide complexes with phosphine oxides are prepared[61] by bromine oxidation of plutonium tribromide in methyl cyanide in the presence of the ligand; this method can also be used to prepare the neptunium tetrabromide complexes.

(l) Nitrogen-donor complexes

Thorium tetrabromide provides a number of examples of nitrogen-donor complexes, whereas the remaining actinide tetrabromides have scarcely been investigated. Most of the known complexes probably involve 8-coordinate thorium, the common stoicheiometry being $ThBr_4 \cdot 4 L$, where L is ethylamine[88], aniline[89] or methyl cyanide[89]. Analogous protactinium[87], uranium[76], neptunium[61] and plutonium[61] tetrabromide complexes with methyl cyanide are known, all of which are isostructural with the thorium compound. Ammines of composition from $ThBr_4 \cdot 8 NH_3$ to $ThBr_4 \cdot 20 NH_3$ appear to exist[90] at 0° and pyridine forms a complex[89] $ThBr_4 \cdot 3 py$. No structural information is available for any of these compounds. They are prepared from the tetrabromide and the ligand, except for the neptunium and plutonium tetrabromide complexes with methyl cyanide, which are made by oxidation of the tribromide with bromine in methyl cyanide[61]. The plutonium compound is rather unstable with respect to decomposition to plutonium(III), as would be expected. In the case of the methyl cyanide adducts, the ligand is held only rather weakly, and this probably applies to all of the compounds mentioned above.

(m) Tetraiodides

Only three tetraiodides are known (Th, Pa, U) and of these the

uranium compound decomposes appreciably to the triiodide when
heated in a vacuum. The possible existence of NpI_4 in a UI_4 matrix has
been mentioned earlier (p. 79). Thorium tetraiodide is of monoclinic
symmetry (space group $P2_1/n-C_{2h}^5$), the thorium atom being surrounded
by 8 iodine atoms in a deformed square antiprismatic array[91]. These
antiprisms share edges and triangular faces to form a layer structure, in
which the layers are weakly bonded to each other. The structures of the
other two tetraiodides are unknown.

Thorium[7] and uranium[92] tetraiodides are best prepared from the
elements at temperatures of the order of 400–500° and with an excess
of iodine present in the case of the uranium compound; protactinium
tetraiodide is most conveniently made by reduction of the pentaiodide
with aluminium metal at 450° in a vacuum[52]. All three are hygroscopic,
moisture-sensitive compounds, soluble in water with appreciable hydrolysis. The melting points are moderately low (ThI_4, 566°; UI_4, in iodine
vapour, 506°) and they can be sublimed in a vacuum (Th, Pa) or in
iodine vapour (to prevent decomposition, U) at moderate temperatures.

(n) Iodocomplexes

Anionic iodocompounds of the form $M_2^IM^{IV}I_6$, in which the
actinide ion is octahedral, are known for thorium[92], protactinium[85] and
uranium[92]; they can be isolated from non-aqueous media such as methyl
cyanide provided that large enough cations (e.g. $(C_4H_9)_4N^+$, $(C_6H_5)_4As^+$)
are used. These complexes are, however, appreciably more difficult to
prepare than their hexachloro- or hexabromo-analogues, the order of
complexing ability of halide ion for the actinides being $F > Cl > Br > I$.
That is, the actinides exhibit Chatt-Ahrland A-class, or hard-acid,
behaviour.

These compounds are soluble in acetone and in methyl cyanide;
they are hydrolysed rapidly by water, and more slowly in moist air, so
that they must be prepared under anhydrous conditions.

(o) Oxygen-donor complexes

Very few complexes of the tetraiodides are known and there are no structural data available for any of them. Complexes with N,N-dimethylacetamide[70], $ThI_4 \cdot 6$ DMA and $UI_4 \cdot 4$ DMA, can be prepared from the components in non-aqueous media and an analogous complex of N,N-dimethylformamide with uranium tetraiodide can apparently[93] be isolated from an aqueous hydriodic acid solution of uranium(IV). This last is said to be stable in air, in contrast to the DMA complexes, which hydrolyse in moist air.

(p) Nitrogen-donor complexes

The thorium complex with methyl cyanide, $ThI_4 \cdot 4$ CH_3CN, can be isolated from solutions of the tetraiodide in the ligand[92], and the protactinium tetraiodide complex, made in a similar manner, probably has the same composition[52]; the uranium tetraiodide complex has not been isolated. Some ammines are known, such as the amidoiodide, $Th(NH_2)_2I_2 \cdot 3$ NH_3, obtained by reaction of the tetraiodide with potassamide in liquid ammonia[94], and the compounds $UI_4 \cdot 10$ NH_3 and $UI_4 \cdot 4–5$ NH_3, obtained from the tetraiodide and gaseous ammonia at 50–100° and 150–200° respectively[232]. Ammonolysis to a nitride iodide occurs above 350°.

(q) Oxohalides

Compounds of the general type MOX_2 (M = Th–Np, X = Cl, Br; M = Th, X = F or I; M = Pa, X = I) are quite well known. The structure of $ThOF_2$ may be LaF_3-type[95], or an orthorhombic cell related to it[96]. The four oxodichlorides are isostructural, and the full structure of $PaOCl_2$ has been determined[97]; this consists of infinite polymeric chains in which the repeating unit is $Pa_3O_3Cl_6$, the three protactinium atoms being respectively 9-coordinate [effectively a tricapped (one Cl, two O) trigonal prism (two Cl and one O on each triangular face)], 8-coordinate (dodecahedral, five Cl, three O) and 7-coordinate (derived from the dodecahedron but with one vertex unoccupied) (Fig. 11). The crystal symmetry is rhombohedral (space group *Pbam*). No other complete structures have been reported.

References p. 133

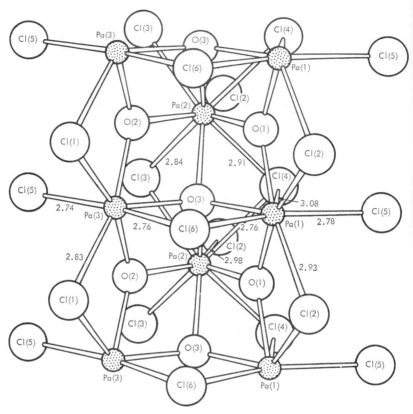

Fig. 11. The atomic arrangement in the polymeric chain in $PaOCl_2$. All atoms are in horizontal mirror planes passing approximately out of the plane of the paper (R.P. Dodge, G.S. Smith, Q. Johnson and R.E. Elson, *Acta Cryst.*, B24 (1968) 304). Reproduced by permission of the International Union of Crystallography.

The chlorides, bromides and thorium oxodiiodide are most easily prepared by heating the tetrahalide with the stoicheiometric quantity of antimony(III) oxide in a vacuum[98]:

$$3 MX_4 + Sb_2O_3 \rightarrow 3 MOX_2 + 2 SbX_3$$

the antimony trihalide sublimes from the reaction mixture. An alternative method is to heat together the stoicheiometric quantities of the tetrahalide and the dioxide (ThOF$_2$ at[96] 900°, UOBr$_2$ at[99] 500°, ThOI$_2$ at[100] 600°):

$$MX_4 + MO_2 \rightleftharpoons 2 MOX_2$$

PaOI$_2$ is[52] formed when the tetraiodide is sublimed in silica, and is iso-structural with ThOI$_2$. Some of the thorium compounds have also been prepared by pyrolysis of the hydrated tetrahalides.

All of these compounds are involatile and disproportionate above 500° in a vacuum. A chlorocomplex, M^{I}ThOCl$_3$, has been described[56], but it requires further investigation.

5. THE QUINQUEVALENT ACTINIDES

Although this valence state is exhibited by the elements from protactinium to americium inclusive, pentahalides (Table 19) are known only for protactinium (F, Cl, Br, I), uranium (F, Cl, Br) and neptunium (NpF$_5$), a reflection of the increasing oxidising power of the quinque-valent elements as the atomic number increases; anionic fluorocom-plexes are, however, known for the elements from protactinium to plutonium, prepared by dry methods.

TABLE 19

THE ACTINIDE PENTAHALIDES[a]

PaF$_5$	PaCl$_5$ (301)	PaBr$_5$	PaI$_5$
white	yellow	dark-red	black
UF$_5$ (348?)	UCl$_5$	UBr$_5$	
bluish-white	brown	brown-red	
NpF$_5$			

[a]Melting points (°C) in parentheses.

(a) Pentafluorides

Uranium pentafluoride is dimorphic, but both forms are of tetra-gonal symmetry; the coordination number of uranium in the high temperature α-form (space group $I4/m$) is 6, the structure consisting of an infinite chain of octahedra linked by opposite corners[101], whereas in the low temperature β-form (space group $I\bar{4}2d$) the uranium atom is 7-coordinate, the arrangement being a pentagonal bipyramid in which four of the vertices of the pentagon are shared[101], so forming an infinite chain of pentagonal bipyramids. Protactinium pentafluoride is isostructural[102] with β-UF$_5$, as is neptunium pentafluoride[103], which is somewhat surprising in view of the smaller ionic radius of neptunium(V).

Compounds of composition intermediate between MF$_4$ and MF$_5$ are definitely established for protactinium and uranium, but are much less certain for plutonium and there appear to be no neptunium analogues. The best known of these compounds are U_2F_9, in which the uranium atoms are probably 9-coordinate[104], and U_4F_{17}, the X-ray powder pattern of which is very similar[104] to that of monoclinic UF$_4$. In the case of protactinium, the only recorded species is isomorphous[2] with U_2F_9, although analysis[42] suggests that its composition is close to Pa_4F_{17}. Pu_4F_{17} has also been reported[105] as present in the residue left on reaction of the tetrafluoride with fluorine, but another attempt to prepare it[106] was unsuccessful.

The uranium intermediate fluorides are prepared by the reaction of uranium tetrafluoride with the hexafluoride under reduced pressure, U_2F_9 being formed at 200° at 17.7 mm pressure of UF$_6$, whereas at 300° at the same pressure U_4F_{17} results, while at 100° and the same pressure of UF$_6$, the product is βUF$_5$. At higher pressures of UF$_6$ (120–140 mm) at 200° the product is αUF$_5$, while at 300° U_2F_9 is formed[107]. The protactinium intermediate fluoride is formed in the reaction of protactinium pentoxide with a mixture of hydrogen and hydrogen fluoride[42]. Protactinium pentafluoride is best prepared by reaction[102] of the tetrafluoride with fluorine at 700°; the corresponding reaction[108] with uranium tetrafluoride yields αUF$_5$ at 150°, and the hexafluoride is formed at higher temperatures. However, fluorine oxidation of UF$_4$ in anhydrous liquid hydrogen fluoride yields βUF$_5$ at room temperature[109] and this last can also be prepared by the reaction of UCl$_5$ or UCl$_6$ with anhydrous liquid hydrogen fluoride at room temperature[108] and by the reaction of UF$_6$ with gaseous thionyl chloride at

room temperature[233]. Alternative methods, such as the reaction of UF_4
with UF_6, mentioned previously, can also be used.

Neptunium pentafluoride[103] has only been obtained by the reduction of the hexafluoride with iodine in iodine pentafluoride at room
temperature; some tetrafluoride is also formed in the reaction.

All of the uranium compounds disproportionate when heated, the
pentafluorides slowly above 150° and the intermediate fluorides at
higher temperatures[107]. Protactinium pentafluoride is appreciably less
volatile than the niobium or tantalum compounds, subliming in vacuum
above 500°, a difference which is probably, at least in part, a result of
the polymeric structure of PaF_5, in contrast to the tetrameric niobium
and tantalum compounds. The pentafluorides are moisture sensitive
and, as would be expected, uranium (and probably neptunium) pentafluoride disproportionate in water. However, protactinium pentafluoride
forms a dihydrate, $PaF_5 \cdot 2 H_2O$, which crystallizes from aqueous hydrofluoric acid[102, 110] and is also formed when protactinium(V) hydroxide
is treated with anhydrous hydrogen fluoride at room temperature. This
last reaction yields[2] the monohydrate at 60°.

(b) Fluorocomplexes

Complexes of the general form $A^I MF_6$ ($Pa^{111, 112}$, $U^{113, 114}$,
Np^{115} and Pu^{116}), $A_2^I MF_7$ ($Pa^{112, 117}$, U^{118}, Np^{115} and Pu^{116}) and
$A_3^I MF_8$ ($Pa^{112, 119}$, $U^{118, 120, 121}$ and Np^{122}) are known; the hexa-
fluoroanions in the lithium (U), sodium (U) and caesium (U, Np and
Pu) salts which have been examined are all of octahedral symmetry, but
in the potassium (Pa, U), ammonium (Pa, U), rubidium (Pa, U) salts and
in $CsPaF_6$, all of which are of orthorhombic symmetry (space group
$Cmca$-D_{2h}^{18}), each metal atom is surrounded by eight fluorine atoms in a
dodecahedral array[123]. These differences are presumably a combined
result of the actinide contraction and the size of the alkali metal cation.
Similar differences have been observed in the structures of the hepta-
fluorocompounds; the rubidium salts (U, Np, Pu) are isostructural with
K_2NbF_7 (monoclinic, space group $P2_1/c$-C_{2h}^5) in which the niobium
atom is 7-coordinate, the arrangement being a trigonal prism capped on
one lateral face[124], whereas protactinium in the $M_2^I PaF_7$ (K, NH_4, Rb,

Cs) is 9-coordinate[125,126], the geometry being close to the tricapped trigonal prismatic arrangement of the UCl_3 structure.

The sodium salts of the octafluorocomplexes (Pa, U, Np) are of tetragonal symmetry (space group $I4/mmm$-D_{4h}^{17}) and a structure determination of the protactinium compound has shown that the metal atom is surrounded by 8 fluorine atoms at the corners of an almost perfect cube[122], an arrangement which requires the participation of f-orbitals in the bonding. The potassium (Pa, U), rubidium (Pa) and caesium (Pa) salts are of face-centred cubic symmetry (space group $Fm3m$-O_h^5) but the coordination geometry is not known.

All three types of complex can be isolated from aqueous hydrofluoric acid solutions of protactinium(V), although not all of the possible alkali metal salts of each type can be prepared in this way; surprisingly, the method applies to even the hexafluorouranates(V) provided that the concentration of hydrofluoric acid is sufficiently high to prevent disproportionation; the hydrated acid, $HUF_6 \cdot 2 H_2O$, has been isolated from concentrated hydrofluoric acid solution[113] and the anhydrous species is apparently formed as a green solid when UF_6 is dissolved in trifluoroacetic acid[237]. The uranium, neptunium and plutonium fluorocomplexes are normally prepared by oxidation of a mixture of alkali metal fluoride and actinide tetrafluoride, in the appropriate proportions for the desired product, with fluorine at moderate temperatures, a method used also for some of the hexafluoroprotactinates(V). Na_3NpF_8 has also been prepared by the reaction of solid sodium fluoride with NpF_6 at[159] 250-400°.

The protactinium compounds appear to be quite stable with respect to hydrolysis, while the uranium(V) compounds disproportionate in water in the absence of hydrofluoric acid (see above) and $CsNpF_6$ disproportionates in anhydrous hydrofluoric acid[127]. It appears, however, that the octafluoroneptunates(V) disproportionate only to a small extent in an aqueous mixture of perchloric and boric acids[128]. On standing, $CsPuF_6$ decomposes[116] to $CsPuF_5$, probably as a result of the plutonium α-radiation.

(c) Pentachlorides

Only two pentachlorides are known, $PaCl_5$ and UCl_5, and these differ markedly in structure; $PaCl_5$ is of monoclinic symmetry (space group $C2/c$ or Cc) and adopts the 7-coordinate, pentagonal bipyramidal arrangement[129] as described for PaF_5 and βUF_5, whereas in UCl_5 (monoclinic, space group $P2_1/n-C_{2h}^5$) the arrangement is octahedral, two octahedra sharing an edge to form the U_2Cl_{10} unit[130]. This dimeric structure persists in carbon tetrachloride solution and even in the vapour phase[234].

The best methods of preparing protactinium pentachloride are either by the reaction of the hydroxide, previously vacuum dried at 100°, with thionyl chloride vapour at 350–500° in a sealed vessel, or by heating a mixture of the pentoxide and carbon with chlorine and carbon tetrachloride at 550° in a sealed tube. Some of the oxochloride, Pa_2OCl_8, is formed in this last reaction[131]. However, uranium pentachloride is much more difficult to obtain in a pure state because it loses chlorine at about 100° forming the tetrachloride, and disproportionates to a mixture of the hexachloride and tetrachloride at a similar temperature. It is probably best prepared by the reaction of UO_3 with $SiCl_4$ at 400° in a sealed tube; the product is soluble in $SiCl_4$, giving a red solution, and is isolated by evaporation of the $SiCl_4$ in a vacuum[137]. Other, less satisfactory, preparative methods are by heating the tetrachloride in a stream of chlorine at 550° and rapidly quenching the vapour and by chlorination of a uranium oxide (UO_3 or U_3O_8) with a mixture of chlorine and carbon tetrachloride in a sealed tube[133] at 250°.

The pentachlorides are very sensitive to moisture; both are hydrolysed and the hydrolysis is accompanied by disproportionation in the case of uranium(V). Solutions in thionyl chloride are easily obtained by refluxing freshly precipitated protactinium(V) hydroxide or uranium trioxide with this solvent. Protactinium pentachloride is also soluble in methyl cyanide, tetrahydrofuran and slightly soluble in carbon tetrachloride; uranium pentachloride can be recrystallised from the last[134] and from liquid chlorine. It is also soluble in carbon disulphide.

(d) Chlorocomplexes

Hexachlorocomplexes (Pa^{135}, U^{136}), in which the anion MCl_6^- is octahedral, and octachlorocomplexes (Pa, U) have been recorded. The hexachlorocompounds $M^IM^VCl_6$ (M^I = NMe_4, Me_2NH_2, Ph_4As) are precipitated from thionyl chloride solutions containing equimolar quantities of the component halides on addition of carbon disulphide[135,136]; the tetrapropylammonium and rubidium[137] salts have also been obtained from thionyl chloride solution, and the caesium salt is precipitated when a solution of caesium chloride in a mixture of thionyl chloride and iodine monochloride is added to a thionyl chloride solution of the actinide pentachloride[135,136]. The alkali metal compounds are more easily prepared by the reaction of a mixture of the alkali metal chloride and UO_3 (or U_3O_8) with thionyl chloride in a sealed tube at 180–200° for 24 h[235]. Other, and much simpler, preparative routes for the tetraalkylammonium salts are by chlorine oxidation of the uranium(IV) complex in a nitromethane solution at[138] 80–90°, and by the reaction of equimolar amounts of the hexachlorouranates(IV) and tetrahalodioxouranates(VI) with an excess of anhydrous liquid hydrogen chloride[138];

$$UCl_6^{2-} + UO_2Cl_4^{2-} + 10\,HCl \rightleftharpoons 2\,UCl_6^- + 4\,HCl_2^- + 2\,H_3O^+$$

Octachlorocomplexes are reported[135,136] to be precipitated by carbon disulphide from thionyl chloride solutions of the stoicheiometric quantities of tetramethylammonium chloride and the actinide(V); the existence of the uranium compound now appears to be somewhat doubtful[138]. All of these compounds are unstable to moisture, rapidly undergoing hydrolysis and disproportionation. The uranium compounds are, however, much more stable to heat than the parent pentachloride.

The compounds $2\,PaCl_5 \cdot SOCl_2$ and $UCl_5 \cdot SOCl_2$, isolated from thionyl chloride solution, and the compound $UCl_5 \cdot PCl_5$, prepared by heating uranium trioxide with phosphorus pentachloride, are probably more correctly formulated as hexachlorometallates, $SO(PaCl_6)_2$, $SOCl(UCl_6)$ and $(PCl_4)(UCl_6)$.

(*e*) *Oxygen-donor complexes*

Phosphine oxide complexes of the types $MCl_5 \cdot R_3PO$(Pa^{139}, U^{140}) and $MCl_5 \cdot 2\ R_3PO$(Pa, U^{141}; $R = C_6H_5$, C_8H_{17}, $(CH_3)_2N$) have been reported, but structural data are not available for them. The protactinium compounds of the first type are prepared by treating a solution of the pentachloride in methyl cyanide or methylene dichloride with the ligand, and also by the reaction of $CsPaCl_6$ with the ligand in methylene dichloride[139]; the last is the method used for the preparation of the analogous uranium complexes[140]. The *bis*-phosphine oxide complexes have been prepared[141] from methyl cyanide or methylene dichloride solutions of the components. The trichloroacryloyl chloride ($CCl_2 = CClCOCl$, TCAC) complex, formed as an intermediate in the chlorination of UO_3 with hexachloropropene ($CCl_3CCl = CCl_2$), is now known[137] to have the composition $UCl_5 \cdot TCAC$. All of these complexes are moisture sensitive and readily hydrolysed, the hydrolysis leading to disproportionation in the case of the uranium compounds. The tri-N,N-dimethylphosphoramide, $[(CH_3)_2N]_3PO$, complex of UCl_5 is only known as a transient species in solution, the ligand causing disproportionation of the uranium(V) even in the absence of water; aryl and alkaryl ketones have a similar effect on uranium pentachloride[140].

(*f*) *Nitrogen-donor complexes*

A methyl cyanide complex, $PaCl_5 \cdot 3\ CH_3CN$, can be isolated from a methyl cyanide solution of protactinium pentachloride[2]; uranium pentachloride will probably behave in a similar manner. Other complexes with pyridine and allied bases, of the form $UCl_5 \cdot 2$-$3L$, are said to be precipitated from a thionyl chloride solution of the components on the addition of petroleum ether (40-60°)[236].

(*g*) *Pentabromides*

Protactinium pentabromide exists in two crystal modifications; high temperature $\beta PaBr_5$ is monoclinic (space group $P2_1/n$) and is isostructural with UCl_5, the compound being a dimer in which two octahedral $PaBr_6$ units share an edge[142]. The low temperature α-form is also of monoclinic symmetry but the structure has not been determined. Uranium pentabromide is isostructural with[143] βPaBr$_5$.

The protactinium compound is prepared either by heating a mixture of the pentoxide and carbon with bromine in a sealed tube at 600–700°, a reaction in which the oxotribromide, $PaOBr_3$, is also formed[144] or, better, by heating protactinium pentachloride with boron tribromide under reflux[82]. The product is sublimed in a vacuum, sublimation at 390–400° yielding the α-form, whereas at 400–410° the β-form is obtained[142]. Uranium pentabromide is somewhat more difficult to prepare; it was reported[145] to be formed by heating uranium trioxide in carbon tetrabromide at 110–130°, a reaction in which careful control of the temperature was said to be necessary in order to avoid the formation of the tetrabromide or the oxotribromide, $UOBr_3$, but this preparation could not be repeated[143]. However, the pentabromide can be prepared by the reaction of uranium metal or the tetrabromide with bromine, either at room temperature in the presence of a small (catalytic) amount of methyl cyanide (which forms a complex $UBr_5 \cdot CH_3CN$) or with the boiling halogen[143].

The pentabromides behave in much the same way as the pentachlorides towards water, and are also soluble in methyl cyanide. Uranium pentabromide is soluble in liquid bromine, which can be used to extract the compound from its reaction mixtures[143]; it dissolves in acetone with decomposition.

(h) Bromocomplexes

Tetraalkylammonium salts, $NR_4MBr_6(Pa$[146], U[138]), in which the anion is octahedral, are prepared either from methyl cyanide solutions of the component halides[146] $[(C_2H_5)_4NPaBr_6]$ or by adding an excess of bromine to a solution of the corresponding uranium(IV) complex in dry nitromethane at room temperature[138] $[(C_2H_5)_4NUBr_6, (C_6H_5)_4 AsUBr_6]$ Alternatively, the hexachlorocomplexes can be converted to the hexabromocompounds (Pa, U) by reaction with boron tribromide[82].

The uranium(V) compounds are also obtained by the reaction of equimolar amounts of the hexabromouranate(IV) and tetrabromodioxouranate(VI) with an excess of anhydrous liquid hydrogen bromide[138]. The acid, $HUBr_6$, appears to be formed when uranium pentabromide is dissolved in a solution of hydrogen bromide in methylene dichloride[143]. These hexabromocomplexes appear to be appreciably more sensitive to moisture than the hexachlorocomplexes, as one might expect from the weaker complexing ability of the bromide ion, as compared to the chloride ion, in the actinide series. The tetraphenylarsonium salts are soluble in methylene dichloride[82].

(i) Oxygen-donor complexes

Complexes of uranium pentabromide with phosphine oxides, $UBr_5 \cdot R_3PO$ [R = C_6H_5, $(CH_3)_2N$], are obtained by the oxidation of uranium tetrabromide by bromine in anhydrous methyl cyanide in the presence of the ligand or by reaction of the corresponding uranium pentachloride complex with boron tribromide, a reaction used also for the preparation[82,141] of $PaBr_5 \cdot 2(C_6H_5)_3PO$. These compounds are soluble in methyl cyanide and in methylene dichloride, and are very moisture sensitive. The uranium compounds also dissolve in acetone with decomposition.

(j) Nitrogen-donor complexes

Methyl cyanide complexes, $MBr_5 \cdot 3\ CH_3CN$, are known; the protactinium compound[144] is obtained from a methyl cyanide solution of the pentabromide and the uranium one[143] by bromine oxidation of uranium tetrabromide in methyl cyanide or of $UBr_4 \cdot 4\ CH_3CN$ in chloroform or methylene dichloride[143]. Very little is known about these compounds.

(k) Pentaiodides

Protactinium pentaiodide is prepared from the elements at 450° or, more conveniently, by reaction of the pentachloride or pentabromide

with silicon tetraiodide at $180°$. It sublimes in a vacuum at $450°$ and is
soluble in methyl cyanide. It is extremely moisture-sensitive[147].

(l) Iodocomplexes

The protactinium compound, $(C_6H_5)_3(CH_3)AsPaI_6$, has been
isolated from a methyl cyanide solution of the component iodides[147].
Although UI_5 is unknown, when hydrogen iodide is condensed on to
salts such as $(C_2H_5)_4NUCl_6$ or $(C_6H_5)_4AsUCl_6$ at $-78°$, a reaction
occurs when the hydrogen iodide melts and a black solid, probably the
hexaiodouranate(V), is formed. This decomposes to uranium(IV) at
$-30°$. A similar product is obtained by reaction of tetrachlorodi-
oxouranates(VI) with anhydrous hydrogen iodide, presumably by way
of deoxygenation of the uranyl, UO_2^{2+}, group[138].

(m) Oxohalides

The known actinide(V) oxohalides are listed in Table 20. The
only structure reported for any of these oxohalides is that of $PaOBr_3$,
for which a 7-coordinate, slightly distorted pentagonal bipyramidal
arrangement has been found[148]; this is achieved by replacing three
halogen atoms in the pentagon by oxygen atoms, each of which is
bonded to two protactinium atoms in the plane of the pentagonal
sheets and to one in an adjacent sheet (Fig. 12).

TABLE 20

ACTINIDE (V) OXOHALIDES

Pa_2OF_8	$NpOF_3 \cdot 2 H_2O$		NpO_2F
U_2OF_8			
Pa_2OCl_8	$PaOCl_3(?)$	$Pa_2O_3Cl_4$	PaO_2Cl
	$UOCl_3$		$UO_2Cl(?)$
			$PuO_2Cl(?)$
	$PaOBr_3$		PaO_2Br
	$UOBr_3$		UO_2Br
	$PaOI_3$		PaO_2I

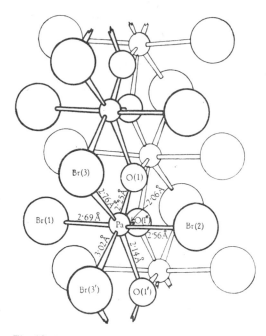

Fig. 12. A perspective view of the double chain in PaOBr$_3$ (D. Brown, T. Petcher and A.J. Smith, *Nature*, 217 (1968) 738). Reprinted by permission of the publishers, Macmillan, London.

Compounds of the type M_2OX_8 are known only for cases where the parent pentahalide exists in the 7-coordinate pentagonal bipyramidal structure; since Pa_2OCl_8 is dimorphic and in view of the structure of PaOBr$_3$, this suggests that their structures could be derived from the pentagonal bipyramidal arrangement of the parent compound by replacing a terminal halogen atom *either* in two pentagons *or* in two axial positions with one oxygen atom. The result would then be linked sheets or parallel chains as shown in Fig. 13.

If the view just expressed is correct, the corresponding oxofluorides should also be dimorphic. Pa_2OF_8 is known to be isostructural with U_2F_9, which is hardly surprising in view of the similar sizes of the O^{2-} and F^- ions, and $NpOF_3$ is isostructural with NpO_2F_2, for the same reason. Pa_2OF_8 is prepared[102] by thermal decomposition of $PaF_5 \cdot 2 H_2O$ at 160°, by the oxidation of PaF_4 with a mixture of fluorine and

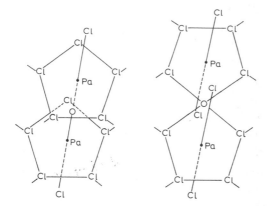

Fig. 13. Possible coordination arrangements in Pa_2OCl_8.

oxygen and by heating the pentoxide in a mixture of hydrogen fluoride
and oxygen. It is also formed by the reaction of $PaOBr_3$ with hydrogen
fluoride at room temperature[2], and decomposes above $800°$, yielding
PaF_5 and other, unidentified, products. The uranium(V) compound is
formed when uranium tetrafluoride is heated at $850°$ in an intermittent
flow of oxygen[149]. It is less stable to heat than Pa_2OF_8, and decomposes in a vacuum at $300°$:

$$2\,U_2OF_8 \;\rightarrow\; UF_6 \;+\; UO_2F_2 \;+\; 2\,UF_4$$

The corresponding protactinium oxochloride, αPa_2OCl_8, has been
isolated as a by-product of the reaction of chlorine and carbon tetrachloride vapour with a mixture of protactinium pentoxide and carbon[131].
It remains when the pentachloride has been removed by sublimation at
temperatures below $220°$ (10^{-4}-mm pressure). The compound
disproportionates above $250°$ in a vacuum:

$$3\,\alpha Pa_2OCl_8 \;\rightarrow\; 4\,PaCl_5 \;+\; Pa_2O_3Cl_4$$

The β-form is obtained when protactinium pentachloride is heated with
a deficit of oxygen in a sealed tube at[131] $350{-}400°$, as a result of the
reaction:

$$4 \, PaCl_5 + O_2 \rightarrow 2 \, \beta Pa_2OCl_8 + 2 \, Cl_2$$

All of these compounds are moisture sensitive.

The only oxofluoride of the type MOX_3 is the neptunium compound. The dihydrate, $NpOF_3 \cdot 2 \, H_2O$, is prepared by heating the pentoxide with hydrogen fluoride at 40°; this is readily dehydrated at 150° in hydrogen fluoride[150]. It is stable towards hydrolysis in moist air. The corresponding chlorides are known for both protactinium and uranium, although the existence of the former is rather uncertain. The uranium compound is formed either by heating together the stoicheiometric quantities of uranium tetrachloride and uranyl chloride at[151] 370° or by the chlorination of uranium dioxide with carbon tetrachloride at 400–500°; a postulated reaction is:

$$2 \, UO_2 + 2 \, CCl_4 \rightarrow 2 \, UOCl_3 + 2 \, CO + Cl_2$$

The protactinium compound appears to be formed by the reaction of protactinium pentachloride with the appropriate quantity of oxygen, but the preparative method seems[131] to give somewhat erratic results in this instance.

The oxotribromides (Pa, U) are quite well established; the protactinium compound is best prepared by heating protactinium pentabromide with the stoicheiometric amount of oxygen[144] or antimony(III) oxide[2] at 350° in a sealed tube; it disproportionates above 500° in a vacuum:

$$2 \, PaOBr_3 \rightarrow PaO_2Br + PaBr_5$$

The uranium compound is prepared[145] by heating the trioxide in a stream of nitrogen and carbon tetrabromide vapour at 110°. It loses bromine slowly at room temperature and forms the uranium(IV) oxodibromide, $UOBr_2$, at 300° in dry nitrogen. It is insoluble in carbon tetrachloride and carbon disulphide, but dissolves in chloroform and bromoform[152].

The only example of an oxotriiodide is the protactinium compound, prepared[147] by heating the pentaiodide with the stoicheiometric quantity of antimony(III) oxide in a vacuum at 150°. Like the oxotribromide, it disproportionates above 450°.

The protactinium oxochloride, $Pa_2O_3Cl_4$, is obtained either by heating Pa_2OCl_8 above 250° (p. 116) or by reaction of the pentachloride with the stoicheiometric quantity of oxygen at[131] 500°:

$$4\ PaCl_5\ +\ 3\ O_2\ \rightarrow\ 2\ Pa_2O_3Cl_4\ +\ 6\ Cl_2$$

It disproportionates above 520° in a vacuum:

$$2\ Pa_2O_3Cl_4\ \rightarrow\ PaCl_5\ +\ 3\ PaO_2Cl$$

The only dioxomonofluoride is the neptunium compound; products of composition close to NpO_2F are obtained by reduction of neptunyl(VI) fluoride, NpO_2F_2, with hydrogen at 325–350°. There appears to be an appreciable range of stoicheiometry for this species[150]. It is unaffected by moist air.

Protactinium dioxomonochloride is formed in the disproportionation of $Pa_2O_3Cl_4$ (see above) and the uranium compound may be formed in the electrolytic reduction of uranyl(VI) chloride in molten salts[153]. There is also spectroscopic evidence for its formation in fused lithium chloride–potassium chloride solutions of uranyl chloride, resulting from decomposition of the latter[151]. The analogous protactinium oxobromide is formed in the disproportionation of the oxotribromide[144] (p. 117), whereas the uranium compound is obtained by heating uranium trioxide with hydrogen bromide at 250°; the hydrogen bromide adduct, $UO_2Br \cdot 2\ HBr$, is formed when uranyl chloride is treated with hydrogen bromide at room temperature. The anhydrous compound decomposes at 500° in nitrogen:

$$2\ UO_2Br\ \rightarrow\ 2\ UO_2\ +\ Br_2$$

The only known dioxomonoiodide is the protactinium compound, a product of the disproportionation of $PaOI_3$; this is also obtained by heating the pentaiodide with the stoicheiometric amount of antimony(III) oxide at 150°.

All of the oxochlorides, oxobromides and oxoiodides described above are sensitive to moisture and are quite readily hydrolysed; their infrared spectra indicate that they are all oxygen-bridged polymeric species but, apart from PaOBr$_3$, no structures are known at present.

(n) Oxohalocomplexes

The dioxodifluorocomplexes, $A^I M^V O_2 F_2$ (A^I = K, NH$_4$, Rb; M = Np[156], Pu[156], Am[156, 157]), are all of rhombohedral symmetry (space group $R3m$-D_{3d}^5) but the full structures are unknown. They are precipitated from aqueous solutions of the actinide(V) on addition of the alkali metal fluoride.

Chlorocomplexes of this composition are unknown, but compounds of the type $Cs_3 M^V O_2 Cl_4$ (M = Np[54], Am[158]) have been reported; the neptunium compound is precipitated when acetone is added to a nearly neutral aqueous solution of neptunium(V) containing caesium chloride. The isostructural americium[158] compound is formed when americium(V) hydroxide is treated with a solution of caesium chloride in concentrated hydrochloric acid.

The remaining oxohalocomplexes are of the type $A_2^I M^V OX_5$; the protactinium oxochloro- and oxobromocomplexes of this composition [A^I = (C$_2$H$_5$)$_4$N] are prepared[141] by treating the tetrahalodiethoxy-complexes, [(C$_2$H$_5$)$_4$N] Pa(OC$_2$H$_5$)$_2$X$_4$, with slightly moist methyl cyanide at 80°.

Analogous uranium(V) complexes, $A_2^I UOX_5$ [X = F, A^I = (C$_2$H$_5$)$_4$N; X = Cl, A^I = (C$_2$H$_5$)$_4$N, (C$_6$H$_5$)$_4$As; X = Br, A^I = (C$_6$H$_5$)$_4$As] have recently been reported[138]. The problem with complexes of this composition is that the analyses could equally well fit equimolar mixtures of $A_2^I M^{IV} Cl_6$ and $A_2^I M^{VI} O_2 Cl_4$, resulting from disproportionation of the actinide(V) species. However the colours, UV/visible and infrared spectra of the above oxohalo-species show that they are indisputably uranium(V) compounds. The blue oxochloro- and green oxobromo-complexes are prepared by limited hydrolysis of the corresponding hexahalouranate(V) in nitromethane in the presence of an excess of the univalent cation halide, preferably at low temperature. Under these conditions the pyridinium hexachlorouranate(V) yields a product of composition $(pyH)_2UOCl_5 \cdot 2.5\,pyHCl$ and $(C_2H_5)_4NUBr_6$ yields the analogous $[(C_2H_5)_4N]_2UOBr_5 \cdot 2.5(C_2H_5)_4NBr$. The pink oxofluorocomplex, $[(C_2H_5)_4N]_2UOF_5$, is obtained as the dihydrate when the corresponding hexachlorouranate(V) is added to a solution of slightly moist tetraethylammonium fluoride in a mixture of acetone and ethanol (9:1) at 0–10°. The product yields the anhydrous compound when vacuum dried.

The corresponding neptunium(V) chlorocompounds, $A_2^INpOCl_5$ $[A^I = Cs, (C_6H_5)_4As]$, obtained[54] by treating freshly precipitated neptunium(V) hydroxide with a saturated solution of the cation chloride in concentrated hydrochloric acid, may well be mixtures of neptunium(IV) and neptunyl(VI) species.

(o) Oxygen–donor complexes

Triphenyl phosphine oxide complexes, $PaOX_3 \cdot 2\,L$ (X = Cl, Br) have been recorded[141]; the compound $UOBr_3 \cdot 2.5\,DMA$ [DMA = $CH_3CON(CH_3)_2$], obtained by the reaction of the amide with uranium pentabromide in acetone, is actually[143] $[UO_2(DMA)_5]^{2+}[UBr_6]^{2-}$.

6. THE SEXIVALENT ACTINIDES

Hexafluorides, MF_6 (U, Np, Pu), and one hexachloride, UCl_6 are the only non-oxygenated actinide(VI) compounds known (Table 21),

TABLE 21

THE ACTINIDE HEXAHALIDES[a]

UF_6 (64)	NpF_6 (55.1)	PuF_6 (51.59)
white	orange	reddish-brown
UCl_6[b]		
blackish-green		

[a]Triple point (°C) in parentheses.
[b]Melting point 177.5°.

but derivatives of the dioxocations, MO_2^{2+}, are found for the four
actinides from uranium to americium inclusive.

(a) Hexafluorides

The solid hexafluorides are of orthorhombic symmetry (space
group $Pnma\text{-}D_{2h}^{16}$) and the molecular geometry is that of a regular
octahedron. They are low melting, volatile solids, as is commonly the
case with both metal and non-metal hexafluorides, in which inter-
molecular interactions are minimal. They are normally prepared by
reaction of the tetrafluorides with fluorine at $220°$ (U), $500°$ (Np) or
$750°$ (Pu). Attempts to prepare the analogous americium compound by
the reaction of the oxide, Am_2O_3, with fluorine in the presence of
platinum, with PtF_6 as a carrier gas, were not successful[160]. Vapour-
pressure data are available for all three hexafluorides[161, 162]; the
volatility of the liquids decreases in the order $UF_6 > NpF_6 > PuF_6$ but
in the solid state NpF_6 has the highest vapour pressure of the three.

The thermal stability also decreases in the order $UF_6 > NpF_6 > PuF_6$
and in the original preparations of[163] NpF_6 and[164] PuF_6 liquid fluorine
was allowed to drip onto the heated lower fluoride, in an apparatus in
which the product hexafluoride could be condensed close to the
reaction area. However, this is an unnecessary complication, for NpF_6
can be obtained in good yield by heating the tetrafluoride in a stream of
fluorine in a tube furnace, the hexafluoride being condensed in a cold
trap beyond the furnace[165].

Uranium hexafluoride is also obtained by reaction of sulphur
tetrafluoride with either uranium trioxide or uranyl fluoride above $300°$:

$$UO_3 + 3 SF_4 \rightarrow UF_6 + 3 SOF_2$$

Uranyl fluoride is formed as an intermediate in this reaction[166]. Above
$500°$, however, uranium hexafluoride is reduced by sulphur tetrafluoride:

$$UF_6 + SF_4 \rightarrow UF_4 + SF_6$$

and plutonium hexafluoride is reduced in the same way at 30°, so providing a method of separating the two elements[167].

All three hexafluorides are immediately hydrolysed by water but they do not react with quartz or Pyrex if water or hydrogen fluoride is absent. If either are present, even in trace amounts, complete hydrolysis occurs by way of cyclic reactions in which hydrogen fluoride reacts with silica:

$$SiO_2 + 4\,HF \rightarrow SiF_4 + 2\,H_2O$$

and is then regenerated:

$$MF_6 + 2\,H_2O \rightarrow MO_2F_2 + 4\,HF$$

Consequently it is advisable to store these compounds over anhydrous sodium or potassium fluoride, which takes up both hydrogen fluoride and water. The hexafluorides are also decomposed, with liberation of free fluorine, by α-radiation. This effect is particularly marked with the plutonium compound, for which the rate of α-radiation decomposition in the solid is about 2% per day[168], but the effects of α-radiation are considerably reduced for the compound in the vapour phase because of the lower probability of a collision between a molecule of the hexafluoride and an α-particle. For this reason it is best to store the hexafluorides as gases at low pressure.

The hexafluorides are powerful fluorinating agents, being reduced by phosphorus, arsenic, silicon, carbon disulphide and most organic compounds, but reduction by hydrogen is slow at room temperature. Uranium hexafluoride is reduced to the quinquevalent state by gaseous nitric oxide or nitrogen dioxide, with the formation of the hexafluorouranates(V) $NOUF_6$ and NO_2UF_6. Plutonium hexafluoride is also known to react with nitric oxide, but the product has not been identified. Uranium hexafluoride is stable in solution in carbon tetrachloride, chloroform and s-tetrachloroethane at room temperature, but reacts with these solvents at higher temperatures.

(b) Fluorocomplexes

Hepta- and octa-fluorouranates(VI), $M^I UF_7$ and $M_2^I UF_8$, are well established, and there seems to be no reason why the analogous neptunium and plutonium compounds should not be obtained in the future. Sodium heptafluorouranate(VI), a white solid, is most conveniently prepared by reaction of uranium hexafluoride with sodium fluoride in n-perfluoroheptane at $100°$; it decomposes in a vacuum at $100°$ to yield yellow $Na_2 UF_8$. The potassium salts are made in the same way[169] and the pale-yellow ammonium salt, $NH_4 UF_7$, is obtained either by adding solid ammonium fluoride to a solution of uranium hexafluoride in chlorine trifluoride[171] or by condensing uranium hexafluoride vapour in a suspension of ammonium fluoride in s-tetrachloroethane[170]. The caesium salt, $CsUF_7$, is precipitated when a solution of caesium fluoride in chlorine trifluoride is added to uranium hexafluoride[171] and when caesium fluoride is heated in UF_6 vapour at 200-$220°$; the yellow hydrazinium salt, $N_2 H_5 UF_7$, crystallises when uranium hexafluoride is mixed with hydrazinium fluoride in anhydrous hydrogen fluoride[172]. The greenish-yellow nitrosonium, and greenish-white nitronium compounds, respectively $NOUF_7$ and $NO_2 UF_7$, are obtained directly from uranium hexafluoride and the corresponding fluoride[173].

X-Ray data[169] indicate that $Na_2 UF_8$ has body-centred tetragonal symmetry (space group $I4/mmm-D_{4h}^{17}$); the structure is, however, incompletely worked out. $CsUF_7$ possesses cubic symmetry and it appears that the fluorine atoms occupy seven-eighths of the positions on the diagonals of a centred cube[238].

(c) Uranium hexachloride

This is the only actinide hexachloride known; it is a blackish-green solid of hexagonal symmetry (space group $C\bar{3}m-D_{3d}^3$) in which the individual UCl_6 molecules are almost perfect octahedra[174]. It begins to decompose, with loss of chlorine, at 120-$150°$, melts at about $178°$ and is relatively volatile, subliming at $100°/10^{-4}$ torr[175].

The best method of preparing the compound is probably by the reaction of the hexafluoride with an excess of boron trichloride or aluminium trichloride, the excess being necessary because uranium hexachloride reacts with the hexafluoride yielding the tetrafluoride and free chlorine[176]. Alternative methods of preparing the compound are by the disproportionation of the pentachloride (p. 109), which is itself difficult to prepare, or, in rather poor yield, by reaction of the tetrachloride with chlorine above 350° and also by the action of a mixture of carbon tetrachloride and chlorine on uranium trioxide at 65-170° under pressure[177,178].

The hexachloride is immediately hydrolysed to uranyl chloride by water, but is soluble in carbon tetrachloride and in chloroform, slightly soluble in n-perfluoroheptane but insoluble in benzene. It reacts with hydrogen fluoride to form UF_5 with the liberation of chlorine, and with uranium hexafluoride, as mentioned above. The yellow precipitate obtained when α,α-dipyridyl is added to solution of uranium hexachloride in carbon tetrachloride may be a nitrogen-donor adduct[134].

(d) Oxohalides

The actinide(VI) oxohalides are all of the form MO_2X_2, and are commonly known as actinyl halides. The uranyl compounds have been very extensively investigated, but very little is known about the analogous neptunyl, plutonyl and americyl halides, the chemical behaviour of which should show some resemblance to that of the uranyl species. The principal difference between these actinyl ions lies in the increasing oxidising power of the MO_2^{2+} ion as one passes along the group from uranium. This is most apparent in the behaviour of the dioxocations with halide ion; AmO_2^{2+} is reduced to americium(V) by chloride or bromide ion in aqueous acid, and to americium(III) by iodide ion, which reduces PuO_2^{2+} and NpO_2^{2+} to plutonium(III) and neptunium(IV) respectively, while even uranyl iodide does not seem to have been isolated as the simple, unsolvated compound.

(e) Dioxodifluorides

The dioxodifluorides (U, Np, Pu, Am) are all of rhombohedral symmetry (space group $R\bar{3}m-D_{3d}^5$); the structural arrangement in the uranium compound consists of a distorted hexagonal bipyramid, with the O=U=O group normal to the median plane of the six fluorine atoms, which are arranged around the uranium atom in a puckered hexagon, the fluorine atoms being alternately above and below the median plane containing the uranium atom. The general shape of the structural unit is rather like the chair form of cyclohexane[179] (Fig. 14).

Fig. 14. Structural arrangement of part of a layer in UO_2F_2. The fluorine atoms Ⓕ lie 0.61 Å above or below the plane of the uranium atoms of the uranyl, UO_2^{2+}, groups, which are shown as ⊕ (After W.H. Zachariasen, *Acta Cryst.*, 1 (1948) 277).

Anhydrous uranyl fluoride is a pale-yellow, hygroscopic solid, prepared by heating uranium tetrafluoride in oxygen[180]:

$$2\,UF_4 + O_2 \rightarrow UF_6 + UO_2F_2$$

and also by heating uranium trioxide in fluorine or with hydrogen fluoride[181] at 350-500°:

$$UO_3 + 2\,HF \rightleftharpoons UO_2F_2 + H_2O$$

The pink neptunyl compound, NpO_2F_2, is prepared in the same way, but from the hydrated trioxide, $NpO_3\cdot H_2O$ at 225° (F_2) or 250-275° (HF), and also by evaporating an aqueous hydrofluoric acid solution of neptunium(VI) to dryness[150]. A white, hydrated plutonyl fluoride is precipitated when a mixture of methanol and hydrofluoric acid is added to an aqueous solution of plutonium(VI); this is dehydrated when washed with anhydrous hydrofluoric acid and dried over phosphorus pentoxide[182]. The pale-brown americyl compound has been made by treating sodium americyl acetate, $NaAmO_2(CH_3COO)_3$, with a mixture of anhydrous hydrogen fluoride and fluorine at $-196°$, followed by removal of the fluorinating mixture under vacuum at room temperature[183].

All of the dioxodifluorides are soluble in water or mineral acids; aqueous solutions of these compounds are obtained when an actinyl(VI) compound (*e.g.* a hydrated trioxide) is dissolved in aqueous hydrofluoric acid and hydrates can be isolated from these solutions. Uranyl fluoride, for example, yields a trihydrate[184] and the compound[185] $UO_2F_2 \cdot 2\,HF \cdot 4\,H_2O$.

The only complexes formed with neutral donor ligands appear to be ammines, such as $UO_2F_2 \cdot 4\,NH_3$, obtained from uranyl fluoride and liquid ammonia[186].

Tri- and penta-fluorocomplexes, such as $CsUO_2F_3$ and $K_3UO_2F_5$, and more complex species of composition $K_3(UO_2)_2F_7$ and $K_5(UO_2)_2F_9$, have been isolated from aqueous solutions; the $UO_2F_5^{2-}$ ion in $K_3UO_2F_5$ is a pentagonal bipyramid[187] and the structures of the other types of fluorocomplex are unknown. The bright green neptunium compound $K_3NpO_2F_5$, isostructural with $K_3UO_2F_5$, is precipitated when a solution of neptunium(IV) or neptunium(V) in saturated potassium fluoride solution is treated with ozone[188]. Plutonium compounds of the types $M^IPuO_2F_3 \cdot H_2O$, $M_2^IPuO_2F_4$ and $Cs(PuO_2)_2F_5 \cdot 3\,H_2O$ are also known[189].

(f) Dioxodichlorides

The only anhydrous dioxodichloride known is the bright-yellow uranyl compound, prepared by heating uranium tetrachloride in oxygen at[190] $300-350°$ and also by heating the hydrates, obtained by evaporating solutions of uranium(VI) in hydrochloric acid to dryness, in hydrogen chloride at $300°$, and then in a mixture of hydrogen chloride

and chlorine at[191] 400°. Anhydrous uranyl chloride melts at 578° and
on ignition in air is converted to the oxide, U_3O_8 ; it decomposes to the
dioxide at 300° in a vacuum and at 400° in nitrogen. It is deliquescent
and is very soluble in water and in many organic solvents[192].

The monohydrate and the trihydrate can be isolated from aqueous
solution; the latter is converted to the monohydrate when treated with
thionyl chloride under reflux[193].

Neptunyl chloride is unknown but the hydrated plutonyl com-
pound, $PuO_2Cl_2 \cdot 6 H_2O$, a greenish-yellow solid, is obtained by vacuum
evaporation of an aqueous hydrochloric acid solution of plutonium(VI)
at room temperature[194]. This compound is hygroscopic and is very
soluble in water; it decomposes slowly, with reduction to plutonium(IV),
presumably a result of the α-radiation of plutonium.

Chlorocomplexes of the type $M_2^I M^{VI}O_2Cl_4$ are known for uranium
(yellow to yellow-green), neptunium (dark yellow), plutonium (yellow)
and americium (dark red). Alkali metal (K, Rb, Cs, NH_4) and tetraalkylammonium salts of the uranium chlorocomplex ion crystallise from
aqueous hydrochloric acid solutions of the appropriate chlorides; the
ammonium, potassium and rubidium compounds obtained in this way
are dihydrates, whereas the salts of the larger cations are anhydrous. The
anhydrous sodium and potassium salts are said to have been prepared
by passing uranyl chloride vapour over the heated alkali metal
chloride[195]. The neptunium[54, 196] and plutonium[196] compounds are
usually isolated from concentrated hydrochloric acid solutions of the
actinide(VI), whereas the americium compound, $Cs_2AmO_2Cl_4$, is obtained by treating the americium(V) chlorocomplex, $Cs_3AmO_2Cl_4$,
with concentrated hydrochloric acid, a reaction which appears to
involve spontaneous oxidation, possibly because of a higher lattice
energy of the americium(VI) compound[158]. It exists in two crystal
modifications, one of which is isostructural with the other $Cs_2MO_2Cl_4$;
the anion is nearly octahedral in the uranium compound, the linear
O-U-O group being almost at right angles to the equatorial plane containing the four chlorine atoms[197].

Dichlorotrioxouranates of the type $M_2^I UO_3Cl_2$ are also known;
the potassium salt has been made by heating potassium uranate, K_2UO_4,

in hydrogen chloride at[198] about 200° and by the reaction of uranyl chloride monohydrate with the stoicheiometric quantity of potassium hydroxide. Since the ammonium salt can also be prepared from uranyl chloride monohydrate, it suggests that the latter could be considered to be the acid $H_2UO_3Cl_2$. Both of these salts react with hydrogen chloride at 150° to form the dioxotetrachlorouranates(VI). Somewhat similar, but substoicheiometric[200], species are obtained by heating uranium trioxide with an alkali metal chloride in a vacuum at 600°.

Oxygen- and nitrogen-donor complexes of uranyl chloride are usually of the type $UO_2Cl_2 \cdot 2$ L (monodentate ligands) or $UO_2Cl_2 \cdot L$ (bidentate ligands). With small ligands, such as urea, the *tetrakis* complex, $UO_2Cl_2 \cdot 4$ L, is obtained[201] but with increasing size, for example with *N*-ethylurea, the *tris* complex, $UO_2Cl_2 \cdot 3$ L, is formed and with tetramethylurea the more usual *bis* complex results[202]. More detailed accounts of these complexes are available in the reviews[1-4]; complexes of the other actinide(VI) dioxodichlorides have not been recorded.

(g) Dioxodibromides

The blood-red uranyl compound is the only known actinide(VI) dioxodibromide. It is very hygroscopic, forming the yellow trihydrate[203], and it decomposes slowly at room temperature with the evolution of bromine. It is difficult to obtain pure, the best methods being by heating the tetrabromide in oxygen at 150–160° (about 96% purity[204]) and by heating $UOBr_3$ in oxygen at 150° (about 98% purity[205]). Evaporation of aqueous solutions of uranyl bromide, obtained by dissolving the trioxide in hydrobromic acid, yields the trihydrate, but a basic compound, $UO_2(OH)Br \cdot 2 H_2O$, crystallises from such solutions if insufficient acid is present[206].

Dioxotetrabromouranates(VI) are obtained from aqueous hydrobromic acid in the same way as the chlorocomplexes; these compounds, which are yellow or yellow-brown, are less stable with respect to reduction than their chloride analogues. The potassium and ammonium salts separate from aqueous hydrobromic acid as the dihydrates and these lose water at $120°$ in nitrogen; the caesium salt separates from aqueous media in the anhydrous condition and an X-ray study has shown that the anion is, like the chloride analogue, almost octahedral[207].

Anhydrous $K_2UO_2Br_4$ reacts with oxygen at $250°$ to form $K_2UO_3Br_2$, also obtained by treating $UO_2Br_2 \cdot H_2O$ with the stoicheiometric quantity of potassium hydroxide, and the corresponding ammonium salt is also known. Both of these react with hydrogen chloride at $150°$ to form[199] salts of the mixed halide complex ion, $[UO_2Cl_2Br_2]^{2-}$. These can also be made[198] by heating uranyl chloride with the alkali bromide at $300°$.

Most of the known oxygen-donor complexes are of the type $UO_2Br_2 \cdot 2$ L, exceptions being the N,N-dimethylformamide (DMF) complex[240], $UO_2Br_2 \cdot 3$ DMF, and the ammines[186], $UO_2Br_2 \cdot 3$ NH_3 and $UO_2Br_2 \cdot 4$ NH_3. No structural data are available for any of these complexes.

(h) Dioxodiiodides

UO_2I_2 is the only compound of this type to be recorded, and this is not known in an unsolvated form. The ether adduct, an orange-red solid which decomposes above $0°$, has been prepared by treating uranyl nitrate trihydrate with barium iodide[55], or uranyl chloride with sodium iodide[208], in both cases in ether, and evaporating the filtrate. It is very hygroscopic, very soluble in water and in many organic solvents.

The only recorded iodocomplex is the dark red $(Ph_3BuP)_2UO_2I_4$, isolated from methyl cyanide solution[209]. N,N-dimethylformamide forms the *tetrakis* complex[210], $UO_2I_2 \cdot 4$ DMF, obtained by adding the ligand to an ethereal solution of the dioxodiiodide, and a *bis*-phosphine oxide adduct is also known[209].

7. THE PSEUDOHALIDES

Unsolvated actinide cyanides and thiocyanates are unknown, and there is very little structural information available for the few complexes which have been recorded. There is one recent review of the known complexes in which the infrared spectral data are summarised[211].

References p. 133

(a) Cyanides

A trichlorocyanotetrammine uranium(IV) complex, $[UCl_3(CN)-(NH_3)_4]$, and a tetrachlorodicyanouranate(IV), $((C_2H_5)_4N)_2[UCl_4(CN)_2]$ are known[212]. The former is precipitated as a light green solid when uranium tetrachloride is treated with sodium or potassium cyanide in anhydrous liquid ammonia, and the latter is prepared by the reaction of uranium tetrachloride with tetraethylammonium cyanide in liquid hydrogen cyanide. Both compounds are very sensitive to atmospheric oxidation and hydrolyse in moist air, indicating that the metal-carbon bond is quite weak[212].

A dioxotetracyanouranate(VI), $K_2[UO_2(CN)_4]$ was reported[55] to be precipitated when a large excess of potassium cyanide was added to an aqueous solution of uranyl(VI) acetate, but this is not a cyanide and is only a product of the hydrolysis of uranyl acetate.

(b) Thiocyanates

A number of oxygen-donor and anionic complexes have been reported for the actinide thiocyanates; the thiocyanate group is usually bonded to the metal in these compounds by the nitrogen atom and, in a few cases, there is evidence for the thiocyanate group acting as a bridge between two actinide metal atoms.

The hydrates, $Th(NCS)_4 \cdot 4 H_2O$, in which thorium is presumably 8-coordinate, and the orange compound $UO_2(NCS)_2 \cdot 3 H_2O$ (probably with uranium 7-coordinate) are usually prepared by treating aqueous thorium(IV) sulphate[213] or uranyl sulphate[214] with barium thiocyanate. Both compounds are extensively dissociated in water, but both dehydrate when heated, with some decomposition, the thorium compound[213] at $100°$ and the uranium one at[215] $115-125°$. The product in the latter case is said[215] to be dark green and insoluble in water, and the original compound may be completely decomposed to, for example, an oxosulphide. The monohydrated uranyl compound, $UO_2(NCS)_2 \cdot H_2O$, prepared by heating the trihydrate at $85-100°$, seems to contain bridging thiocyanate groups, for the C–N stretching mode appears at 2103 cm^{-1}

in the infrared spectrum as compared with 2057 cm^{-1} (terminal NCS) in the trihydrate[216].

A few amide and phosphine oxide complexes of the actinide(IV) thiocyanates are known, in nearly all of which the metal atom appears to be 8-coordinate. Examples are the N,N-dimethylacetamide (DMA) complexes [M(NCS)$_4$(DMA)$_4$](Th[217], U[218]), and the phosphine oxide complexes [M(NCS)$_4$(R$_3$PO)$_4$](Th, U, Np[219]). These compounds are prepared by treating the corresponding complex of the actinide tetra-chloride with potassium thiocyanate in a non-aqueous solvent, usually in the presence of an excess of the ligand. A complex with diphenyl sulphoxide (DPSO), Th(NCS)$_4$·4 DPSO, is also known[243]. It is prepared from the hydrated tetrathiocyanate and the ligand in acetone or by mixing hot acetone solutions of thorium perchlorate, ammonium thiocyanate and the ligand.

Urea complexes of uranyl thiocyanate, such as the yellow [UO$_2$(NCS)$_2$(H$_2$O)(CO(NH$_2$)$_2$)$_2$] and the dark orange [UO$_2$(NCS)$_2$-(CO(NH$_2$)$_2$)$_3$] have been isolated from methanol solutions of the trihydrate on addition of urea[220] and an analogous acetamide complex, [UO$_2$(NCS)$_2$(H$_2$O)(CH$_3$CONH$_2$)$_2$], is obtained when the amide is added to an aqueous solution of uranyl thiocyanate; this loses water[220] at 150–160°. Most of the recorded uranyl thiocyanate complexes are of the general form [UO$_2$(NCS)$_2$L$_3$] where L is antipyrine, tetrahydrofuran, dimethyl sulphoxide[220] or quinoline N-oxide[221]; these are obtained by reaction of the trihydrate with the ligand and, like the latter, are probably examples of 7-coordinate uranium. However, triphenyl phos-phine oxide and the analogous arsine oxide precipitate the *bis* complexes when an ethanolic solution of the ligand is added to the filtrate from the reaction between hydrated uranyl nitrate and an excess of potassium thiocyanate in ethanol[222]. A *bis*-tri-*n*-butyl phosphate (TBP) adduct is also known[216]; in this compound both bridging and terminal thiocyanate groups are present (2110 and 2036 cm^{-1}) but in the presence of an excess of the ligand the band due to bridging thiocyanate groups dis-appears and the *tris* complex [UO$_2$(NCS)$_2$(TBP)$_3$] is formed[216]. The analogous *tris*-tri-*n*-butyl phosphine oxide complex is also known[223].

Anionic actinide(IV) thiocyanato complexes are all of the type

$M_4^I M^{IV}(NCS)_8$ (Th, U). They are prepared by reaction of thorium(IV) or
uranium(IV) sulphate with the stoicheiometric quantities of the alkali
metal (or ammonium) sulphate and barium thiocyanate in sulphuric
acid. Infrared data on the series are available[224]. The tetraethylammonium salt is obtained by reaction of the hexachlorouranate(IV) with
potassium thiocyanate in methyl cyanide, the excess potassium ion
being eliminated by treatment with tetraethylammonium bromide[215]. In
this last the coordination geometry is an almost perfect cube[226]. The
thorium, neptunium and plutonium analogues have the same structure[248].

Uranyl compounds of the types[227] $M^I[UO_2(NCS)_3(H_2O)_2]$ and[228]
$M_3^I UO_2(NCS)_5$ are obtained in the same way as the octa-N-thiocyanato
actinide(IV) complexes. The anion in the latter type of complex is a
pentagonal bipyramid, with five NCS groups in the equatorial plane[229].

(c) Cyanates

Uranyl cyanate, $UO_2(CNO)_2$, and anionic complexes derived from
this compound, have been recorded[239], but only the preparative details
are available for these compounds; it is uncertain whether these involve
nitrogen bonding of the cyanate group to the metal or not.

Yellow $K_2 UO_2(CNO)_2$ is precipitated when a solution of potassium cyanate in aqueous ethanol is added dropwise to a solution of
uranyl nitrate in the same medium. It is very soluble in water, but
dissociation occurs and yellow orange $K(UO_2)_2(CNO)_5$ is precipitated. The
simple cyanate, $UO_2(CNO)_2$, is apparently obtained as a golden yellow
precipitate by treating $K_2 UO_2(CNO)_4$ with uranyl nitrate.

(d) Selenocyanates

A thorium N-selenocyanate complex with N,N-dimethylformamide
(DMF), Th(NCSe)_4·4 DMF, and salts of the solvated anionic complexes,
$Na_2 Th(NCSe)_6$·3 DMF, $K_2 Th(NCSe)_6$·4.5 DMF and $K_4 Th(NCSe)_8$·2 DMF,
are obtained by treating thorium(IV) compounds with the appropriate quantities of an alkali metal (Na, K) selenocyanate in DMF. They
are yellowish solids which decompose in air or in water with the libera-

tion of selenium, and they are evidently much less stable in these respects than are the corresponding thiocyanates. The SeCN groups are monodentate and are bonded to the thorium atom by way of the nitrogen atom[244].

8. REFERENCES

1   K.W. Bagnall in V. Gutmann (Ed.), *Halogen Chemistry*, Vol. 3, Academic Press, London, 1967, p. 303.
2   D. Brown, *Halides of the Lanthanides and Actinides*, Wiley and Sons, London, 1968.
3   D. Brown, *Intern. Rev. Sci.*, 7 (1972) 87, Medical and Technical Publishing Co., and Butterworth, London.
4   K.W. Bagnall, *Coord. Chem. Rev.*, 2 (1967) 145.
5   F. Lux, F. Ammentorp-Schmidt, D. Dempf, D. Graw and W. Hagenberg, *Radiochim. Acta*, 14 (1970) 57.
6   R.J. Clark and J.D. Corbett, *Inorg. Chem.*, 2 (1963) 460.
7   D.E. Scaife and A.W. Wylie, *J. Chem. Soc.*, (1964) 5450.
8   N. Edelstein and W. Easley, *J. Chem. Phys.*, 48 (1968) 2110.
9   N. Edelstein, J.G. Conway, D. Fujita, W. Kolbe and R. McLaughlin, *J. Chem. Phys.*, 52 (1970) 6425.
10 J.R. Peterson and B.B. Cunningham, *J. Inorg. Nucl. Chem.*, 30 (1968) 1775.
11 A. Zalkin, D.H. Templeton and T.E. Hopkins, *Inorg. Chem.*, 5 (1966) 1466.
12 A. Zalkin and D.H. Templeton, *J. Am. Chem. Soc.*, 75 (1953) 2453.
13 C.M. Schwarz and D.A. Vaughan, *U.S. Report BMI-266* (1953).
14 S. Fried and N.R. Davidson, *J. Am. Chem. Soc.*, 70 (1948) 3539.
15 A.E. Florin in G.T. Seaborg, J.J. Katz and W.M. Manning (Eds.), *The Chemistry of the Transuranium Elements*, Natl. Nucl. Energy Ser., Vol. IV–14B, McGraw-Hill, New York, 1949, p. 691.
16 J.J. Katz and E. Rabinowitch, *The Chemistry of Uranium*, Natl. Nucl. Energy Ser., Vol. VIII–5, McGraw-Hill, New York, 1951, pp. 541–542.
17 J.H. Burns, *Inorg. Chem.*, 4 (1965) 881.
18 R.E. Thoma, H.A. Friedman and R.A. Penneman, *J. Am. Chem. Soc.*, 88 (1966) 2046.
19 D.K. Fujita, B.B. Cunningham and T.C. Parsons, *Inorg. Nucl. Chem. Letters*, 5 (1969) 307.
20 Ref. 2, p. 155.
21 S.A. Shchukarev, I.V. Vasil'kova and A.I. Efimov, *Zhur. Neorgan. Khim.*, 1 (1956) 2652.
22 H.J. Berthold and H. Knecht, *Angew. Chem.*, 77 (1965) 428.
23 H.J. Berthold, *Z. Anorg. Allgem. Chem.*, 348 (1966) 50.

24   I. Sheft and S. Fried, *J. Am. Chem. Soc.*, 75 (1953) 1236.
25   K.W. Bagnall, J.B. Laidler and M.A.A. Stewart, *J. Chem. Soc. (A)*, (1968) 133.
26   L.R. Morss and J. Fuger, *Inorg. Chem.*, 8 (1969) 1433.
27   J.L. Ryan, in P.R. Fields and T. Moeller (Eds.), *Lanthanide/Actinide Chemistry*, Adv. Chem. Ser. 71, Am. Chem. Soc., Washington, D.C., 1967, p. 331.
28   Y. Marcus and M. Shiloh, *Israel J. Chem.*, 7 (1969) 37.
29   R. Barnard, J. Bullock and L.F. Larkworthy, *Chem. Commun.*, (1968) 960.
30   R.E. Stevens, *J. Inorg. Nucl. Chem.*, 27 (1965) 1873.
31   S. Fried, D. Cohen, S. Siegel and B. Tani, *Inorg. Nucl. Chem. Letters*, 4 (1968) 495.
32   D. Brown, S. Fletcher and D.G. Holah, *J. Chem. Soc. (A)*, (1968) 1889.
33   Ref. 2, p. 198.
34   V. Scherrer, F. Weigel and M. van Ghemen, *Inorg. Nucl. Chem. Letters*, 3 (1967) 589.
35   N.W. Gregory, *U.S. At. Energy Comm. Rept. TID-5290*, 1958, 465.
36   D. Brown, Personal communication.
37   L.B. Asprey, T.K. Keenan and F.H. Kruse, *Inorg. Chem.*, 3 (1964) 1137.
38   F. Hagemann, B.M. Abraham, N.R. Davidson, J.J. Katz and I. Sheft, in G.T. Seaborg, J.J. Katz and W.M. Manning (Eds.), *The Chemistry of the Transuranium Elements*, Natl. Nucl. Energy Ser., Vol. IV–14B, McGraw-Hill, New York, 1949, p. 957.
39   D. Cohen, S. Fried, S. Siegel and B. Tani, *Inorg. Nucl. Chem. Letters*, 4 (1968) 257.
40   L.B. Asprey, T.K. Keenan and F.H. Kruse, *Inorg. Chem.*, 4 (1965) 985.
41   A.C. Larson, R.C. Roof Jr. and D.T. Cromer, *Acta Cryst.*, 17 (1964) 555.
42   L. Stein, *Proc. Colloq. Intern. Physicochim. Protactinium, Orsay*, 1965, p. 101.
43   I.V. Tananaev and G.S. Savchenko, *Atomniya Energiya*, 12 (1962) 392; *Soviet J. At. Energy*, 12 (1962) 418.
44   R.E. Thoma, *Inorg. Chem.*, 1 (1962) 220.
45   J.H. Burns, R.D. Ellison and H.A. Levy, *Acta Cryst.*, B24 (1968) 230.
46   W.H. Zachariasen, *Acta Cryst.*, 1 (1948) 265.
47   G. Brunton, *Acta Cryst.*, 26B (1970) 1185.
48   D. Brown, in Gmelin, *The Transplutonium Elements*, Vol. 71, Chapters 3.5–3.8 (1972).
49   D.M. Gruen and C.W. de Kock, *J. Inorg. Nucl. Chem.*, 29 (1967) 2569.
50   K. Mucker, G.S. Smith, Q. Johnson and R.E. Elson, *Acta Cryst.*, 25B (1969) 2362.
51   E.R. Harrison, *U.K. At. Energy Authority, Rept.* (1958) A.E.R.E. GP/R 2409.
52   D. Brown and P.J. Jones, *J. Chem. Soc. (A)*, (1967) 719.
53   J.A. Herman and J.F. Suttle, *Inorganic Syntheses*, Vol. 5, McGraw-Hill, New York, 1957, p. 145.

54   K.W. Bagnall and J.B. Laidler, *J. Chem. Soc.* (A), (1966) 516.
55   J. Aloy, *Ann. Chim. Phys.*, 24 (1901) 412.
56   E. Chauvenet, *Compt. Rend.*, 148 (1909) 1519; *Ann. Chim. Phys.*, 23 (1911) 425.
57   R. Benz and R.M. Douglas, *J. Inorg. Nucl. Chem.*, 23 (1961) 134.
58   P.-T. Clève, *Bull. Soc. Chim. Fr.*, 21 (1874) 115.
59   B.W. Fitzsimmons, P. Gans, B.C. Smith and M.A. Wassef, *Chem. Ind., London*, (1965) 1698.
60   P. Day and L.M. Venanzi, *J. Chem. Soc.* (A), (1966) 197.
61   D. Brown, J. Hill and C.E.F. Rickard, *J. Chem. Soc. (A)*, (1970) 497.
62   K.W. Bagnall, D. Brown, P.J. Jones and J.G.H. du Preez, *J. Chem. Soc. (A)*, (1966) 737.
63   J.G.H. du Preez and F.G. Sadie, *J. S.African Chem. Inst.*, 19 (1966) 73.
64   Z.M.S. Al-Kazzaz, K.W. Bagnall and D. Brown, to be published.
65   K.W. Bagnall, D. Brown, P.J. Jones and P.S. Robinson, *J. Chem. Soc.*, (1964) 2531.
66   K.W. Bagnall, D. Brown, F. Lux and G. Wirth, *Z. Naturforsch.*, 24b (1969) 214.
67   K.W. Bagnall, A.M. Deane, T.L. Markin, P.S. Robinson and M.A.A. Stewart, *J. Chem. Soc.*, (1961) 1611.
68   K.W. Bagnall, D. Brown, D.G. Holah and F. Lux, *J. Chem. Soc. (A)*, (1968) 465.
69   K.W. Bagnall, J.G.H. du Preez and M.L. Gibson, *J. Chem. Soc. (A)*, (1971) 2124.
70   K.W. Bagnall, D. Brown, P.J. Jones and J.G.H. du Preez, *J. Chem. Soc.*, (1965) 3594.
71   H.C.E. Mannerskantz, G.W. Parshall and G. Wilkinson, *J. Chem. Soc.*, (1963) 3163.
72   J.M. Matthews, *J. Am. Chem. Soc.*, 20 (1898) 815.
73   H.J. Berthold and H. Knecht, *Angew. Chem.*, 77 (1965) 453.
74   I. Kalnins and G. Gibson, *J. Inorg. Nucl. Chem.*, 7 (1958) 55.
75   J. Selbin, M. Schober and J.D. Ortego, *J. Inorg. Nucl. Chem.*, 28 (1966) 1385.
76   K.W. Bagnall, D. Brown and P.J. Jones, *J. Chem. Soc.*, (1966) 1763.
77   A. Feltz, *Z. Chem.*, 6 (1966) 318.
78   R. Perrot and C. Devin, *Compt. Rend.*, 246 (1958) 772.
79   R.W.M. D'Eye, *J. Chem. Soc.*, (1950) 2764.
80   D.E. Scaife, *Inorg. Chem.*, 5 (1965) 162.
81   R.M. Douglass and E. Staritzky, *Analyt. Chem.*, 29 (1957) 459.
82   D. Brown, J. Hill and C.E.F. Rickard, *J. Chem. Soc. (A)*, (1970) 476.
83   J. Prigent, *Ann. Chim. (Paris)*, 5 (1960) 65.
84   D. Brown, *J. Chem. Soc. (A)*, (1966) 766.
85   D. Brown and P.J. Jones, *J. Chem. Soc. (A)*, (1967) 243.
86   J.L. Ryan and C.K. Jørgensen, *Mol. Phys.*, 7 (1963) 17.
87   D. Brown and P.J. Jones, *Chem. Commun.*, (1966) 280.

88   J.M. Matthews, *J. Am. Chem. Soc.*, 20 (1898) 839.
89   R.C. Young, *J. Am. Chem. Soc.*, 56 (1934) 29.
90   R.C. Young, *J. Am. Chem. Soc,*, 57 (1935) 997.
91   A. Zalkin, J.D. Forrester and D.H. Templeton, *Inorg. Chem.*, 3 (1964) 639.
92   K.W. Bagnall, D. Brown, P.J. Jones and J.G.H. du Preez, *J. Chem. Soc.*, (1965) 350.
93   M. Lamisse, R. Heimburger and R. Rohmer, *Compt. Rend.*, 258 (1964) 2078.
94   G.W. Watt and S.C. Malhotra, *J. Inorg. Nucl. Chem.*, 11 (1959) 255.
95   W.H. Zachariasen, *Acta Cryst.*, 2 (1949) 388.
96   R.W.M. D'Eye, *J. Chem. Soc.*, (1958) 196.
97   R.P. Dodge, G.S. Smith, Q. Johnson and R.E. Elson, *Acta Cryst.*, B24 (1968) 304.
98   K.W. Bagnall, D. Brown and J.F. Easey, *J. Chem. Soc. (A)*, (1968) 288.
99   N.W. Gregory, *U.S. At. Energy Comm. Rept. TID-5290*, (1958) 498.
100 D.E. Scaife, A.G. Turnbull and A.W. Wylie, *J. Chem. Soc.*, (1965) 1432.
101 W.H. Zachariasen, *Acta Cryst.*, 2 (1949) 296.
102 L. Stein, *Inorg. Chem.*, 3 (1964) 995.
103 J. Holloway, Personal communication.
104 W.H. Zachariasen, *Acta Cryst.*, 2 (1949) 390.
105 J.K. Dawson, R.M. Elliott, R. Hurst and A.E. Truswell, *J. Chem. Soc.*, (1954) 558.
106 B. Weinstock and J.G. Malm, *J. Inorg. Nucl. Chem.*, 2 (1956) 380.
107 P. Agron, *U.S. At. Energy Comm. Rept. TID-5290*, (1958) 610.
108 P. Agron, A. Grenall, R. Kurin and S. Weller, ref. 107, p. 652.
109 L.B. Asprey and R.A. Penneman, *J. Am. Chem. Soc.*, 80 (1967) 172.
110 A. von Grosse, *Proc. Roy. Soc.*, A150 (1935) 365.
111 L.B. Asprey, F.H. Kruse and R.A. Penneman, *J. Am. Chem. Soc.*, 87 (1965) 3518.
112 D. Brown and J.F. Easey, *J. Chem. Soc.(A)*, (1966) 254.
113 L.B. Asprey and R.A. Penneman, *Inorg. Chem.*, 3 (1964) 727.
114 G.D. Sturgeon, R.A. Penneman, F.H. Kruse and L.B. Asprey, *Inorg. Chem.*, 4 (1965) 748.
115 L.B. Asprey, T.K. Kennan and G.D. Sturgeon, *Inorg. Nucl. Chem. Letters*, 2 (1966) 19.
116 R.A. Penneman, G.D. Sturgeon, L.B. Asprey and F.H. Kruse, *J. Am. Chem. Soc.*, 87 (1965) 5803.
117 A. von Grosse, *Proc. Roy. Soc.*, A150 (1935) 353.
118 R.A. Penneman, G.D. Sturgeon and L.B. Asprey, *Inorg. Chem.*, 3 (1964) 126.
119 L.B. Asprey, F.H. Kruse, A. Rosenzweig and R.A. Penneman, *Inorg. Chem.*, 5 (1966) 659.
120 W. Rüdorff and H. Leutner, *Ann. Chem.*, 632 (1960) 1.
121 R. Bougon and P. Plurien, *Compt. Rend.*, 260 (1965) 4217.
122 D. Brown, J.F. Easey and C.E.F. Rickard, *J. Chem. Soc. (A)*, (1969) 1161.
123 J.H. Burns, H.A. Levy and O.L. Keller Jr., *U.S. Rept. ORNL-4146* (1967).

124   G.M. Brown and L.A. Walker, *Acta Cryst.*, 20 (1966) 220.
125   D. Brown and A.J. Smith, *Chem. Commun.*, (1965) 554.
126   D. Brown, J.F. Easey and D.G. Holah, *J. Chem. Soc. (A)*, (1967) 1979.
127   L.B. Asprey and R.A. Penneman, *J. Am. Chem. Soc.*, 89 (1967) 172.
128   J.F. Easey, *Ph. D. thesis*, Leicester University, 1968.
129   R.P. Dodge, G.S. Smith, Q. Johnson and R.E. Elson, *Acta Cryst.*, 22
       (1967) 85.
130   G.S. Smith, Q. Johnson and R.E. Elson, *Acta Cryst.*, 22 (1967) 300.
131   D. Brown and P.J. Jones, *J. Chem. Soc. (A)*, (1966) 874.
132   A.D. Webb, *U.S. Rept. RL-4.6.87* and *RL-4.6.102* (1943), according to
       J.J. Katz and E. Rabinowitch, *The Chemistry of Uranium*, Nat. Nucl.
       Energy Ser., Vol. VIII–5, McGraw-Hill, New York, 1951, p. 489.
133   A. Michael and A. Murphy, *Am. Chem. J.*, 44 (1910) 365.
134   P. Gans, *Ph. D. thesis*, London University, 1964.
135   K.W. Bagnall and D. Brown, *J. Chem. Soc.*, (1964) 3021.
136   K.W. Bagnall, D. Brown and J.G.H. du Preez, *J. Chem. Soc.*, (1964) 2603.
137   J. Selbin, J.D. Ortego and G. Gritzner, *Inorg. Chem.*, 7 (1968) 976.
138   J.L. Ryan, *J. Inorg. Nucl. Chem.*, 33 (1971) 153.
139   D. Brown, J.F. Easey and J.G.H. du Preez, *J. Chem. Soc. (A)*, (1966) 258.
140   K.W. Bagnall, D. Brown and J.G.H. du Preez, *J. Chem. Soc.*, (1965) 5217.
141   D. Brown and C.E.F. Rickard, *J. Chem. Soc. (A)*, (1971) 81.
142   D. Brown, T.J. Petcher and A.J. Smith, *Acta Cryst.*, B25 (1969) 178.
143   F. Lux, G. Wirth and K.W. Bagnall, *Chem. Ber.*, 103 (1970) 2807.
144   D. Brown and P.J. Jones, *J. Chem. Soc. (A)*, (1966) 262.
145   J. Prigent, *Ann. Chim. Fr.*, 5 (1960) 65.
146   D. Brown and P.J. Jones, *J. Chem. Soc. (A)*, (1967) 247.
147   D. Brown, J.F. Easey and P.J. Jones, *J. Chem. Soc. (A)*, (1967) 1698.
148   D. Brown, T. Petcher and A.J. Smith, *Nature*, 217 (1968) 738.
149   S.S. Kirslis, T.S. McMillan and H.A. Bernhardt, *U.S. Rept. K-567* (1950).
150   K.W. Bagnall, D. Brown and J.F. Easey, *J. Chem. Soc. (A)*, (1968) 2223.
151   M.D. Adams, D.A. Wenz and R.K. Steunenberg, *J. Phys. Chem.*, 67
       (1963) 1939.
152   G. Kaufmann and R. Rohmer, *Bull. Soc. Chim. France*, (1961) 1969.
153   R.S. Wilks, *J. Nucl. Materials*, 7 (1962) 157.
154   J.-C. Levett, *Compt. Rend.*, 260 (1965) 4775.
155   L.J. Guggenberger and R.A. Jacobson, *Inorg. Chem.*, 7 (1968) 2257.
156   T.K. Keenan, *Inorg. Chem.*, 4 (1965) 1500.
157   L.B. Asprey, F.H. Ellinger and W.H. Zachariasen, *J. Am. Chem. Soc.*, 76
       (1954) 5235.
158   K.W. Bagnall, J.B. Laidler and M.A.A. Stewart, *Chem. Commun.*, (1967)
       24; *J. Chem. Soc. (A)*, (1968) 133.
159   L.E. Trevorrow, T.J. Gerding and M.J. Steindler, *Inorg. Chem.*, 7 (1968)
       2226.
160   S. Tsujimura, D. Cohen, C.L. Chernick and B. Weinstock. *J. Inorg. Nucl.
       Chem.*, 25 (1963) 226.

138   REFERENCES. 7

161   G.D. Oliver, H.T. Milton and J.W. Grisard, *J. Am. Chem. Soc.*, 75 (1953) 2827.
162   B. Weinstock, E.E. Weaver and J.G. Malm, *J. Inorg. Nucl. Chem.*, 11 (1959) 104.
163   J.G. Malm, B. Weinstock and E.E. Weaver, *J. Phys. Chem.*, 62 (1958) 1506.
164   B. Weinstock and J.G. Malm, *J. Inorg. Nucl. Chem.*, 2 (1956) 380.
165   L.E. Trevorrow, J.J. Gerding and M.J. Steindler, *J. Inorg. Nucl. Chem.*, 30 (1968) 2671.
166   A.L. Oppegard, W.C. Smith, E.L. Muetterties and V.A. Engelhardt, *J. Am. Chem. Soc.*, 82 (1960) 3835.
167   M.J. Steindler, *U.S. Patent 3,046,089* (1962).
168   M.J. Steindler, *U.S. Rept. ANL-6753* (1963).
169   J.G. Malm, H. Selig and S. Siegel, *Inorg. Chem.*, 5 (1966) 130.
170   B. Volavsek, *Croat. Chem. Acta*, 33 (1961) 181.
171   N.S. Nikolaev and V.F. Sukhoverkhof, *Dokl. Akad. Nauk SSSR*, 136 (1961) 621.
172   B. Frlec, B.S. Brčić and J. Slivnik, *Croat. Chem. Acta*, 36 (1964) 173.
173   J.R. Geichman, E.A. Smith and P.R. Ogle, *Inorg. Chem.*, 2 (1963) 1012.
174   W.H. Zachariasen, *Acta Cryst.*, 1 (1948) 285.
175   D. Johnson and T. Butler, *U.S. Rep. CC-1524* (1944, declassified 1964).
176   T.A. O'Donnell, D.F. Stewart and P. Wilson, *Inorg. Chem.*, 5 (1966) 1438.
177   H.G. Reiber, *U.S. Patent* 2,499,836 (1950).
178   R.E. van Dyke and E.C. Ewers, *U.S. Patent* 2,725,279 (1955).
179   M. Atoji and M.J. McDermott, *Acta Cryst.*, 26B(1970) 1540.
180   L.M. Ferris, *J. Am. Chem. Soc.*, 79 (1957) 5419.
181   H.R. Hoekstra, *Inorg. Chem.*, 2 (1963) 492.
182   I.F. Alenchikova, L.L. Zaitseva, L.V. Lipis, N.S. Nikolaev, V.V. Fomin and N.T. Chebotarev, *Zhur. Neorgan. Khim.*, 3 (1958) 951.
183   T.K. Keenan, *Inorg. Nucl. Chem. Letters*, 4 (1968) 381.
184   L.H. Brooks, E.V. Garner and E. Whitehead, *U.K. At. Energy Authority Rept. IGR-TN/CA-277* (1956).
185   Yu.A. Buslaev, N.S. Nikolaev and I.V. Tananaev, *Dokl. Akad. Nauk SSSR*, 148 (1963) 832.
186   A. von Unruh, *Dissert.*, Rostock (1909).
187   W.H. Zachariasen, *Acta Cryst.*, 7 (1954) 783.
188   C.E. Thalmayer and D. Cohen, in P.R. Fields and T. Moeller (Eds.), *Actinide/Lanthanide Chemistry*, Advan. in Chem. Ser. 71, Am. Chem. Soc., Washington, 1967, p. 256.
189   I.F. Alenchikova, L.L. Zaitseva, L.V. Lipis, N.S. Nikolaev, V.V. Fomin and N.T. Chebotarev. *Zhur. Neorgan. Khim.*, 6 (1961) 1513; *Russ. J. Inorg. Chem.*, 6 (1961) 777.
190   O. Johnson, T. Butler and A.S. Newton, *U.S. Rept. TID-5290*, (1958), Book I, p. 1.
191   D.C. Bradley, A.K. Chatterjee and A.K. Chatterjee, *J. Inorg. Nucl. Chem.*, 12 (1959) 71.

192   Reference 16; page 577.
193   J.D. Hefley, D.M. Mathews and E.S. Amis, *Inorganic Syntheses*, Vol. 7,
      McGraw-Hill, New York, 1963, p. 146.
194   I.F. Alenchikova, L.L. Zaitseva, L.V. Lipis and V.V. Fomin, *Zhur.*
      Neorgan. Khim., 4 (1959) 961; *Russ. J. Inorg. Chem.*,4 (1959) 435.
195   J. Aloy, *Bull. Soc. Chim. France*, 25 (1901) 153.
196   J.L. Ryan, *Inorg. Chem.*, 2 (1963) 348.
197   D. Hall, A.D. Rae and T.N. Waters, *Acta Cryst.*, 20 (1966) 160.
198   J. Lucas, *Rev. Chim. Min.*, 1 (1964) 479.
199   J. Prigent and J. Lucas, *Compt. Rend.*, 253 (1961) 474.
200   J.G. Allpress and A.D. Wadsley, *Acta Cryst.*, 17 (1964) 41.
201   V.P. Markov, and I.V. Tsapkina, *Zhur. Neorgan. Khim.*, 4 (1959) 2255;
      Russ. J. Inorg. Chem., 4 (1959) 1028.
202   A. Deptula, *Nukleonika*, 10 (1965) 765.
203   S.A. Shchukarev, I.V. Vasil'kova, V.M. Drozdova and E.E. Frantseva,
      Zhur. Neorgan. Khim., 4 (1959) 39; *Russ. J. Inorg. Chem.*, 4 (1959) 15.
204   F.H. Spedding, A.S. Newton, R. Nottorf, J. Powell and V. Calkins, *U.S.*
      Rept. TID-5290, (1958) Book 1, p. 91.
205   J. Prigent, *Ann. Chim. Fr.*, 5 (1960) 65.
206   S. Peterson, *J. Inorg. Nucl. Chem.*, 17 (1961) 135.
207   Yu.N. Mikhailov, V.G. Kuznetsov and E.S. Kovaleva, *J. Struct. Chem.*
      USSR, 6 (1965) 752.
208   M. Lamisse and R. Rohmer, *Bull. Soc. Chim. France*, (1963) 24.
209   J.P. Day and L.M. Venanzi, *J. Chem. Soc. (A)*, (1966) 1363.
210   M. Lamisse, R. Heimburger and R. Rohmer, *Compt. Rend.*, 258 (1964)
      2078.
211   K.W. Bagnall, *Intern. Rev. Sci.*, 7 (1972) 139; Medical and Technical
      Publishing Co., and Butterworth, London.
212   K.W. Bagnall and J.L. Baptista, *J. Inorg. Nucl. Chem.*, 32 (1970) 2283.
213   A.K. Molodkin and G.A. Skotnikova, *Zhur. Neorgan. Khim.*, 9 (1964) 60;
      Russ. J. Inorg. Chem., 9 (1964) 32.
214   P.·Pascal, *Compt. Rend.*, 158 (1914) 1672.
215   V.P. Markov and E.N. Traggeim, *Zhur. Neorgan. Khim.*, 5 (1960) 1467;
      Russ. J. Inorg. Chem., 5 (1960) 712.
216   V.M. Vdovenko, A.I. Skoblo and D.N. Suglobov, *Zhur. Neorgan. Khim.*,
      13 (1968) 3059; *Russ. J. Inorg. Chem.*, 13 (1968) 1577.
217   K.W. Bagnall, D. Brown, P.J. Jones and P.S. Robinson, *J. Chem. Soc.*,
      (1964) 2531.
218   K.W. Bagnall, D. Brown and R. Colton, *J. Chem. Soc.*, (1964) 2527.
219   Z.M.S. Al-Kazzaz, K.W. Bagnall and D. Brown (1972), to be
      published.
220   R.N. Schelokov, I.M. Shul'gina and I.I. Chernyaev, *Zhur. Neorgan. Khim.*,
      12 (1967) 1246; *Russ. J. Inorg. Chem.*, 12 (1967) 660.
221   A.K. Majumdar and R.G. Bhattacharya, *Science and Culture (Calcutta)*,
      35 (1969) 271.

222   F.A. Hart and J.E. Newbery, *J. Inorg. Nucl. Chem.*, 28 (1966) 1334.
223   S.M. Sinitsyna and N.M. Sinitsyn, *Dokl. Akad. Nauk SSSR*, 168 (1966) 110; *Doklady Chem.*, 166–8 (1966) 467.
224   I.E. Grey and P.W. Smith, *Austral. J. Chem.*, 22 (1969) 311.
225   P. Gans and J.W. Marriage, Personal communication.
226   R. Countryman and W.S. McDonald, *J. Inorg. Nucl. Chem.*, 33 (1971) 2213.
227   V.P. Markov and E.N. Traggeim, *Zhur. Neorgan. Khim.*, 5 (1960) 1493; *Russ. J. Inorg. Chem.*, 5 (1960) 724.
228   V.P. Markov, E.N. Traggeim and I.M. Shul'gina, *Zhur. Neorgan. Khim.*, 9 (1964) 550; *Russ. J. Inorg. Chem.*, 9 (1964) 305.
229   E.G. Arutyunyan and M.A. Porai-Koshits, *Zhur. Strukt. Khim.*, 4 (1963) 110; *J. Struct. Chem. (USSR)*, 4 (1963) 96.
230   H.A. Friedman, C.F. Weaver and W.R. Grimes, *J. Inorg. Nucl. Chem.*, 32 (1970) 3131.
231   J.H. Burns and J.R. Peterson, *Inorg. Chem.*, 10 (1971) 147.
232   W. Burk, *Z. Chem.*, 9 (1969) 233.
233   B. Moncela and J. Kirkindai, *Compt. Rend.*, C267 (1968) 1485.
234   D.M. Gruen and R.L. McBeth, *Inorg. Chem.*, 8 (1969) 2625.
235   E. Stumpp, *Naturwissenschaften*, 56 (1969) 370.
236   R.C. Paul, G. Singh and M. Singh, *J. Inorg. Nucl. Chem.*, 33 (1971) 713.
237   A.T. Sadikova, N.S. Nikolaev and T.A. Rasskazova, *Zhur. Neorgan. Khim.*, 15 (1970) 2010; *Russ. J. Inorg. Chem.*, 15 (1970) 1039.
238   A.T. Sadikova, G.G. Sadikov and N.S. Nikolaev, *Atomniya Energiya*, 26 (1969) 283.
239   P. Pascal, *Bull. Soc. Chim. Fr.*, 15 (1914) 11.
240   G. Kaufman, R. Weiss and R. Rohmer, *Bull. Soc. Chim. Fr.*, (1963) 1140.
241   H.J. Berthold and H. Knecht, *Z. Anorg. Allgem. Chem.*, 356 (1968) 151.
242   J. MacCordick and C. Brun, *Compt. Rend.*, 270C (1970) 620.
243   V.V. Savant and C.C. Patel, *J. Less-Common Metals*, 24 (1971) 459.
244   A.M. Golub and V.A. Kalibabchuk, *Russ. J. Inorg. Chem.*, 12 (1967) 1249.
245   F.H. Kruse, *J. Inorg. Nucl. Chem.*, 33 (1971) 1625.
246   R. Barnard, J.I. Bullock, B.J. Gellatly and L.F. Larkworthy, unpublished work (1971).
247   J.R. Peterson and R.D. Baybarz, *Inorg. Nucl. Chem. Letters*, 8 (1972) 423.
248   Z.M.S. Al-Kazzaz, K.W. Bagnall, D. Brown and B. Whittaker, to be published.

Chapter 8

COMPOUNDS FORMED WITH INORGANIC OXO-ACIDS

1. INTRODUCTION

Most of the information available about actinide compounds of this type relates to thorium(IV) and uranyl(VI) compounds. Much less is known about the analogous species formed by the remaining actinides up to, and including, americium, and beyond this element only scattered information concerning very few compounds has been recorded. In many cases our knowledge of the chemistry of these species is restricted to the analytical composition and method of preparation of the compounds, and few structural studies have been made. The nitrates are probably the best known compounds of this group, largely because of the importance of nitrate systems in the separation chemistry of the actinides.

The compounds discussed in this chapter are described in the order in which the main group element of the anion appears in the Periodic Table, followed by those compounds in which the anion is based on a *d* transition element. Carboxylates and thiocarboxylates are dealt with in Chapter 9.

2. BORATES

Boric acid and alkali borates yield a white precipitate with aqueous thorium(IV) solutions, but the composition of the material is uncertain. The precipitate redissolves in aqueous thorium(IV) nitrate solution and appears to form a sol; when this is coagulated after dialysis, the product has the composition[1] $4 ThO_2 \cdot Th_3(BO_3)_4$. No other borates are known.

3. CARBONATES

(a) Actinide(III) compounds

Americium(III) and curium(III) carbonates are the only compounds reported for the terpositive elements. Pink $Am_2(CO_3)_3 \cdot 2 H_2O$ is formed by the hydrolysis of the trichloroacetate in aqueous solution[2] and this compound, when treated with 0.5 M NaHCO$_3$ or 1.5 M Na$_2$CO$_3$ yields the carbonatocomplexes Na[Am(CO$_3$)$_2$]·4 H$_2$O and Na$_3$[Am(CO$_3$)$_3$]·3 H$_2$O respectively[3]. All three compounds decompose to leave AmO$_2$ when heated[3]. A white curium(III) carbonate is precipitated when an aqueous solution of the tervalent element is neutralised with potassium carbonate[4]; it is probably hydrated, like the americium compound. Both compounds redissolve to an appreciable extent in an excess of aqueous alkali metal carbonate.

(b) Actinide(IV) compounds

Although U(CO$_3$)$_2$ and Pu(CO$_3$)$_2$ may be intermediates in the thermal decomposition of hydrated actinide(IV) oxalates[5,204], the dicarbonates have not otherwise been isolated. A white, basic thorium(IV) compound, ThO(CO$_3$)·2 H$_2$O, is obtained by reaction of hydrated thorium dioxide with carbon dioxide at 30–40 atm pressure and the corresponding octahydrate is precipitated by sodium carbonate from aqueous solutions of quadrivalent thorium. The latter loses six molecules of water in a vacuum to yield the dihydrate, which decomposes above 50° to mixed oxide/carbonate species[6]. The analogous dark-green uranium(IV) compound is precipitated on the addition of ammonium carbonate to a solution containing quadrivalent uranium; it oxidises to hydrated uranyl(VI) carbonate on exposure to dry air[7]. The greenish-blue plutonium(IV) compound, PuO(CO$_3$)·2 H$_2$O, is the end product of the α-radiation decomposition of plutonium(IV) oxalate, Pu(C$_2$O$_4$)$_2$· 6 H$_2$O, and of (NH$_4$)$_6$Pu(C$_2$O$_4$)$_5$ ·x H$_2$O when these compounds are stored for 18 months[188].

Anionic carbonatocomplexes derived from the unknown dicarbonates are, however, well established. A green calcium salt, $Ca_2U(CO_3)_4 \cdot 10\ H_2O$, occurs in nature as the mineral liebigite which is identical with another mineral, uranothallite; the crystals are of orthorhombic symmetry[8]. The analogous guanidine ammonium tetracarbonatouranate(IV) is known[8a], but the more common species are the pentacarbonatocomplexes, $[M(CO_3)_5]^{6-}$, known for thorium[9-11], uranium[12] and plutonium[10,13] (Table 22). However, alkali metal tetracarbonatoplutonates(IV), $M_4^I Pu(CO_3)_4 \cdot x\ H_2O$ (M^I = Na, K) are known and potassium salts of the hexa- and octa-

TABLE 22

ACTINIDE(IV) CARBONATOCOMPLEXES ISOLATED FROM AQUEOUS SOLUTION

	Colour	Reference
$(NH_4)_2 Th(CO_3)_3 \cdot 6\ H_2O$	white	9
$Na_6 Th(CO_3)_5 \cdot 12\ H_2O$	white	9
$K_6 Th(CO_3)_5 \cdot 10\ H_2O$	white	9
$[Co(NH_3)_6]_2 [Th(CO_3)_5 H_2O] \cdot 9\ H_2O^a$	orange	11
$[CN_3H_6]_6 U(CO_3)_5 \cdot 4\ H_2O^b$	grey-green	12
$[Co(NH_3)_6]_2 Pu(CO_3)_5 \cdot H_2O] \cdot 4\ H_2O$	(?)	10
$M_6^I Pu(CO_3)_5 \cdot n\ H_2O^c$	green	13

[a]The trihydrate is also known[10].
[b]CN_3H_6 = guanidinium. Thorium(IV) compound also known.
[c]M^I = Na, K or NH_4.

carbonatoplutonates(IV) have been reported[13]. These compounds are precipitated from aqueous carbonate solutions of the quadrivalent actinide by the appropriate cation carbonate or are prepared by variations of this procedure, such as the treatment of powdered uranium(IV) disulphate with aqueous guanidine carbonate saturated with carbon dioxide[12]. Guanidine salts of the basic carbonates, $(CN_3H_6)_5 [M(CO_3)_3(OH)_3] \cdot 5\ H_2O$ (M = Th, U), are said[12] to be formed under similar conditions at 70°. Structural data are not available for these compounds and it is by no means certain whether these species are true pentacarbonatocomplexes or whether, for example, the alkali metal compounds are of the form M_4^I [Th(CO$_3$)$_4$]·M$_2^I$CO$_3$·12 H$_2$O (M^I = Na, K)[14].

(c) Actinide(V) compounds

Actinide(V) carbonatocomplexes, M^I[M^{V}O$_2$CO$_3$], have been reported for neptunium (M^I = K[15]), plutonium (M^I = K[15,16], NH$_4$[16]) and americium (M^I = K[15], NH$_4$, Rb[16]). They are described as light-coloured solids, precipitated from solutions of the alkali metal carbonate and the quinquevalent actinide (Np, Pu) or on the oxidation of americium(III) in alkali carbonate solution at 80° by the hypochlorite of the same alkali metal. The pale-grey trihydrated ammonium salt, NH$_4$PuO$_2$CO$_3$·3 H$_2$O, is also known[17]. The potassium salts (Np, Pu, Am) are isostructural[15], as are the potassium, ammonium and rubidium americium(V) complexes; all of them possess hexagonal symmetry (space group $C6/mmc$–D$_{6h}^4$) and the structure consists of endless layers of composition [M^{V}O$_2$CO$_3$]$^-$ held together by alkali metal cations between the layers. Each M^V atom forms six bonds to the carbonate oxygen atoms in addition to the linear O=M=O group normal to the plane containing the carbonate oxygen atoms[18], the overall geometry of the M^V environment being approximately hexagonal bipyramidal (Fig. 15). The pale greenish-blue neptunium(V) tricarbonatocomplexes, M_5^I [NpO$_2$(CO$_3$)$_3$], (M^I = K, Cs) are obtained by treating neptunium(V) hydroxide with a large excess of 50% aqueous potassium or caesium carbonate[19]. Other actinide(V) carbonatocomplexes, which possess orthorhombic or monoclinic symmetry, have been reported but their stoicheiometry is uncertain[16]. It is probable that these include the tricarbonatocomplexes, M_5^I [M^{V}O$_2$(CO$_3$)$_3$]; the americium(V) complex of this composition appears to be coprecipitated with the uranyl(VI) carbonatocomplex, K$_4$[UO$_2$(CO$_3$)$_3$][19].

(d) Actinide(VI) compounds

Anhydrous uranyl carbonate, UO$_2$CO$_3$, is a creamy-white solid; it cannot be isolated from aqueous media owing to hydrolysis, and is

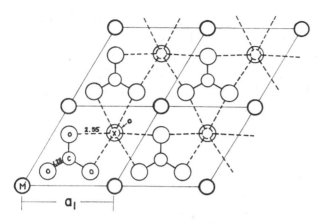

Fig. 15. The projection of a half-cell of $M^I XO_2 CO_3$ on the basal plane (F.H. Ellinger and W.H. Zachariasen, *J. Phys. Chem.*, 58 (1954) 405). Copyright (1954) by the American Chemical Society and reprinted by permission of the copyright holder.

normally prepared[20-22] by heating uranium trioxide in water with carbon dioxide under pressure (15 000 psi) at 300°. The yellow hydrate, $UO_2CO_3 \cdot 2.5\ H_2O$, results when the uranium(IV) basic carbonate is exposed to air[7]. It dehydrates at 160° and the anhydrous product is stable to 500°. The synthetic carbonate is isomorphous[23] with the naturally occurring compound, the mineral rutherfordine, which possesses[24] orthorhombic symmetry (space group $Pmmn$-D_{2h}^{13}). The uranium atom is 8-coordinate in the compound, the geometrical arrangement being close to a hexagonal bipyramid[24]. The rose-red plutonyl(VI) compound, PuO_2CO_3, is obtained by heating the carbonatocomplex, $(NH_4)_4PuO_2(CO_3)_3$, at 120–130°; it decomposes at 130–140° to plutonium dioxide[189].

Actinide(VI) carbonatocomplexes are usually of the form $M_4^I[M^{VI}O_2(CO_3)_3]$, where M^{VI} is U[11,25,26] (yellow), Np[201] (yellow-green), or Pu[19,189] (green). They are prepared by treating uranium trioxide with sodium hydrogen carbonate in water[26], or by crystallisation from concentrated aqueous ammonium[11] or sodium[25] carbonate containing the actinide(VI), or when solid ammonium carbonate is added to a plutonium(VI) solution[189]. Other complexes, such as $Na_6[(UO_2)_2(CO_3)_5]$ and $Na_6[(UO_3)_3(UO_2)_2(CO_3)_5]$, are

References p. 177

obtained by using the appropriate quantities of uranium trioxide and sodium hydrogen carbonate[26]; a basic plutonium(VI) compound, $NH_4[PuO_2(CO_3)(OH)(H_2O)_3]$, is precipitated as a pink solid when ethanol is added to a saturated solution of ammonium diplutonate (probably $PuO_2(OH)_2$) in aqueous ammonium carbonate[190].

4. SILICATES

Actinide silicates are less well known than the carbonates and are of little importance; hydrated thorium silicate, $ThSiO_4 \cdot x\, H_2O$, occurs in nature as the mineral thorite, and the germanate is also known. The thorium, and other actinide(IV) silicates can be prepared by heating the appropriate dioxide with silicon dioxide. A uranyl(VI) silicate, prepared by the action of aqueous uranyl fluoride on silicon dioxide at 300° under pressure, may be non-stoicheiometric[27], the composition being between $(UO_2)_2SiO_4 \cdot 3\, H_2O$ and $(UO_2)_3Si_2O_7 \cdot 3\, H_2O$.

5. NITRATES

Nitrates of the terpositive actinides are unknown; this is hardly suprising for the earlier actinides, in view of the ease of their oxidation to higher states, but such compounds should be obtainable for americium and the following elements in the actinide group.

(a) Actinide(IV) compounds

The only anhydrous tetranitrate known is the thorium compound; treatment of a solution of thorium nitrate tetrahydrate in anhydrous nitric acid with dinitrogen pentoxide yields the compound $Th(NO_3)_4 \cdot 2\, N_2O_5$ which loses N_2O_5 at $160°/10^{-5}$ torr to give the unsolvated tetranitrate. This is very soluble in water and very slightly soluble in diethyl ether[28].

Hydrates of thorium tetranitrate with 5 to 12 molecules of water are obtained from aqueous nitric acid solutions of hydrated thorium dioxide under varying conditions; these can be dried to yield hydrates with from 1 to 4 molecules of water, depending upon the conditions of drying. An investigation of the thorium nitrate–water system showed[29] that the most stable hydrates are $Th(NO_3)_4 \cdot 4 H_2O$ and $Th(NO_3)_4 \cdot 5 H_2O$. The crystal structure of the pentahydrate has been determined by X-ray[30] and neutron[31] diffraction techniques. The crystals are of orthorhombic symmetry (space group $Fdd2$) and the thorium atom is 11-coordinate, being bonded to 8 oxygen atoms from the four bidentate nitrate groups and to the oxygen atoms of three water molecules. The remaining two molecules of water are not bonded to the metal. The coordination arrangement is highly unsymmetrical (Fig. 16), but if the nitrate groups are regarded as single units, the geometry is approximately that of a singly capped trigonal prism. Hydrolysis of aqueous thorium nitrate leads to the formation of the dinuclear species $[Th_2(OH)_2(NO_3)_6(H_2O)_6] \cdot 2 H_2O$. The crystal symmetry is monoclinic (space group $P2_1/c$); the thorium atoms are joined by two hydroxo-bridges, and each metal atom is 11-coordinate, made up of 6 oxygen atoms from 3 bidentate nitrate groups, 3 oxygen atoms from bonded molecules of water and the 2 oxygen atoms of the bridges. The polyhedron can be described as a somewhat distorted dodecahedron if each nitrate group is considered as a single unit[32] (Fig. 17).

Uranium tetranitrate has not been recorded, although a basic nitrate, $UO(NO_3)_2$, and its tetrahydrate have been reported to exist. These preparations could not be repeated[33] and, since oxidation to uranium(VI) occurs readily in uranium(IV)–nitric acid systems, it is unlikely that the simple nitrates will ever be isolated.

The reaction of neptunium tetrachloride with dinitrogen pentoxide yields the off-white solvate, $Np(NO_3)_4 \cdot \sim 1.2 N_2O_5$, which decomposes, with partial oxidation to neptunium(V) when warmed[34]. Vacuum evaporation of a solution of neptunium(IV), in dilute (0.1 M) nitric acid yields the grey dihydrate, $Np(NO_3)_4 \cdot 2 H_2O$, which decomposes to NpO_2 above $50°/10^{-3}$ torr[34]. Slow evaporation of a solution of plutonium(IV) in nitric acid, however, yields[35] the pentahydrate, $Pu(NO_3)_4 \cdot 5 H_2O$, which appears to be isomorphous with the analogous

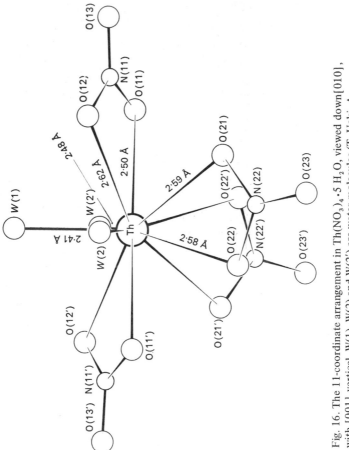

Fig. 16. The 11-coordinate arrangement in Th(NO$_3$)$_4$·5 H$_2$O, viewed down[010], with [001] vertical. W(1), W(2) and W(2') are water molecules (T. Ueki, A. Zalkin and D.H. Templeton, *Acta Cryst.*, 20 (1966) 836). Reproduced by permission of the International Union of Crystallography.

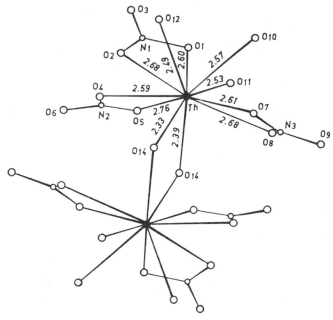

Fig. 17. The atomic arrangement in $[Th_2(OH)_2(NO_3)_6(H_2O)_6]$ as viewed down [100], with [010] vertical (G. Johansson, *Acta Chem. Scand.*, 22 (1968) 389). Reprinted by permission of the publishers.

thorium compound[36]. The plutonium compound is green when finely divided, but large crystals are almost black; it melts at 95–100° and decomposes rapidly at 100°, forming plutonium(VI) species. Its solutions in acetone and in ether are green[35].

Anionic pentanitratocomplexes, $M^I M^{IV}(NO_3)_5 \cdot x\ H_2O$, where x is 9 (M^I = Na, K) or 5 (M^I = NH$_4$) are known only for thorium. They are prepared by evaporating equimolar quantities of the alkali metal nitrate and thorium tetranitrate in neutral or acid solution[37]. The thorium hexanitratocomplexes, $M_2^I Th(NO_3)_6$ (M^I = Na, K, NH$_4$, Rb, Cs) and $M^{II} Th(NO_3)_6 \cdot 8\ H_2O$ (M^{II} = Mg, Mn, Co, Ni, Zn) are prepared in the same way, but from more concentrated nitric acid solutions of the components. The rubidium and caesium salts are much less soluble in nitric acid than are those of the lighter alkali metals[37].

Crystals of $MgTh(NO_3)_6 \cdot 8 H_2O$ have monoclinic symmetry (space group $P2_1/c$) and the compound is actually $[Mg(H_2O)_6] [Th(NO_3)_6] \cdot 2 H_2O$. The other salts, $M^{II}Th(NO_3)_6 \cdot 8 H_2O$ are isomorphous with the magnesium salt. The thorium atom in the anion is 12-coordinate, all six nitrate groups being bidentate, and the coordination polyhedron is an irregular icosahedron[38].

Tetraethylammonium salts of actinide(IV) hexanitratocomplexes are known for thorium[39], uranium[33,40], neptunium[39] and plutonium[39]; they are precipitated from 8 M HNO_3(Th, Np, Pu) or 8 to 15.8 M HNO_3(U) solutions of the components. In the case of the uranium compound, which is almost white, the precipitation is carried out at $0°$ and the solution is saturated with sulphamic acid in order to prevent oxidation by nitrite ion. The caesium salt, $Cs_2U(NO_3)_6$, which is also almost white, is prepared[33] in the same way from 8 M HNO_3. The tetraethylammonium salt (U) can be recrystallised from methanol or acetone, but both of the uranium(IV) nitratocomplexes are unstable to heat and decompose rapidly above about $150°$, with oxidation of uranium(IV) to uranium[33](VI).

A variety of oxygen-donor complexes of the actinide tetranitrates have been prepared in recent years; a list of most of the known compounds is given in Table 23. Dimethyl sulphoxide forms a number of other complexes with thorium tetranitrate in which water is also present[46] and a few Schiff's base[47] and orthophenanthroline[44] adducts of thorium tetranitrate have also been reported.

TABLE 23

SOME OXYGEN-DONOR COMPLEXES OF THE ACTINIDE TETRANITRATES

Ligand	Stoicheiometry	Metals (colour of complex)
$CH_3CON(CH_3)_2$ (DMA)	2 $M(NO_3)_4 \cdot 5$ DMA	Th[41] (white), U[33] (pale-green), Np[34] (grey-green)
$(CH_3)_2SO$ (DMSO)	$M(NO_3)_4 \cdot 6$ DMSO	Th[42] (white), Np[43] (pale-green)
	$M(NO_3)_4 \cdot 3$ DMSO	Th[43] (white), U[43] (dark-green), Np[43] (grey-green)
$(C_2H_5)_2SO$ (DESO)	$M(NO_3)_4 \cdot 3$ DESO	Th[43] (white), Np[43] (brown)
$(C_6H_5)_2SO$ (DPSO)	$M(NO_3)_4 \cdot 3$ DPSO	Th[44,195] (white)
$(C_6H_5)_3PO$ (TPPO)	$M(NO_3)_4 \cdot 2$ TPPO	Th[44] (white), Pu[45] (green)
$(C_6H_5)_3AsO$ (TPAO)	$M(NO_3)_4 \cdot 2$ TPAO	Pu[45] (pink)

The compounds listed in Table 23 are generally prepared by treating the hydrated tetranitrate (Th) or the caesium hexanitratocomplex (Np sulphoxide complexes) with the ligand and either evaporating the resulting solution to dryness or precipitating the complex by adding a hydrocarbon to the solution. The usual method of preparing the uranium(IV) compounds is by metathesis of the corresponding chloride complex with silver nitrate in a dry, non-aqueous solvent such as methyl cyanide.

The only structural data for these complexes refer to $Th(NO_3)_4 \cdot 2\ TPPO$; this is isostructural with the cerium(IV) analogue, which possesses monoclinic symmetry[48] (space group $P2_1/n$). The metal atom is 10-coordinate, and the structure is based on an octahedron, with four bidentate nitrate groups at the corners of the equatorial plane. These nitrate groups are inclined such that the arrangement in the equatorial plane corresponds to that of a four-bladed propeller. The oxygen atoms of the two *trans* TPPO groups are not collinear with the metal atom, the O–M–O angle being $155°$ (Fig. 18).

(b) Actinide(V) compounds

Protactinium(V) and neptunium(V) nitrates have been reported; although no americium(V) nitrates are known, the stability of americium(V) in aqueous solution suggests that at least the hydrate $AmO_2NO_3 \cdot x\ H_2O$ should be preparable.

The hydrated oxotrinitrates, white[49] $PaO(NO_3)_3 \cdot x\ H_2O$ $(1 < x < 4)$ and pink[34] $NpO(NO_3)_3 \cdot 3\ H_2O$, are obtained respectively by treating a protactinium pentahalide with fuming nitric acid and by evaporating a solution of neptunium(V) in 1 M nitric acid to dryness. Evaporation of a neutral solution of neptunium(V) in nitrate media yields the green compound $NpO_2NO_3 \cdot H_2O$. This begins to decompose at $80°/10^{-3}$ torr, but the anhydrous compound can apparently be isolated by heating at $140-220°$. These compounds deliquesce in moist air. A methyl cyanide adduct, $Pa_2O(NO_3)_8 \cdot 2\ CH_3CN$, results from the reaction of protactinium(V) halides in methyl cyanide with dinitrogen tetroxide. Both of the protactinium compounds decompose above $50°$ in a vacuum[49].

References p. 177

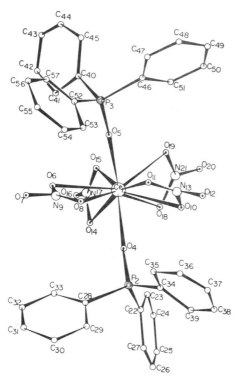

Fig. 18. The structure of tetranitrato*bis*-(triphenylphosphine oxide) cerium(IV).
The thorium analogue has the same structure (Mazur-ul-Haque, C.N. Caughlan,
F.A. Hart and R. van Nice, *Inorg. Chem.*, 10 (1971) 115). Copyright (1971) by the
American Chemical Society and reprinted by permission of the copyright holder.

The reaction of protactinium pentahalides, or of freshly
precipitated protactinium(V) hydroxide, with liquid dinitrogen
pentoxide yields a white compound of composition $HPa(NO_3)_6$ which
is presumably the parent acid for the compounds $M^IPa(NO_3)_6$ [M^I = Cs,
$(CH_3)_4N$, $(C_2H_5)_4N$]. These are white solids prepared by treating the
corresponding hexachlorometallates(V), M^IPaCl_6, with liquid dinitrogen
pentoxide at room temperature. They are soluble in the reagent and are
isolated by evaporation of the solvent[49]. The acid complex decomposes
above 50° in a vacuum[49].

(c) Actinide(VI) compounds

Anhydrous uranyl nitrate is obtained by heating the dinitrogen tetroxide adduct, $UO_2(NO_3)_2 \cdot N_2O_4$, at 163–165°/$10^{-5}$ torr[50–52] or at 163° under atmospheric pressure[53]. It is an orange-yellow solid which is decomposed by diethyl ether at room temperature but which appears to dissolve in the ether without reaction at −60°C. The N_2O_4 adduct, a lemon-yellow solid, is prepared by reaction[51,52] of anhydrous uranium oxides (UO_3, U_3O_8) with liquid N_2O_4 or of the metal[53] with a solution of the reagent in nitromethane. Addition of a large excess of the reagent to the resulting solution precipitates the complex. The reaction of the metal with N_2O_4 in methyl cyanide or ethyl acetate yields yellow $UO_2(NO_3)_2 \cdot 2\ CH_3CN$ (m.p. 180°) and bright-yellow $UO_2(NO_3)_2 \cdot 2\ CH_3COOC_2H_5$ (m.p. 94°) respectively[53]. Hydrated uranium oxides ($UO_4 \cdot 2\ H_2O$, $UO_3 \cdot 2\ H_2O$ or $UO_3 \cdot H_2O$) react with liquid N_2O_4 to yield hydrated dinitrogen tetroxide adducts of uranyl nitrate. These lose the oxide of nitrogen in dry air, forming hydrated uranyl nitrate[51].

The bright yellow hexahydrate, $UO_2(NO_3)_2 \cdot 6\ H_2O$, which melts[54] at 60.3°, is prepared by dissolving uranium trioxide in nitric acid and evaporating the solution until crystallisation begins. It loses water to form lower (di- and tri-) hydrates when heated and ultimately decomposes to uranium trioxide; the anhydrous nitrate cannot be prepared by heating any of the known hydrates.

The pink, hygroscopic neptunyl(VI) nitrate, $NpO_2(NO_3)_2 \cdot 1$–2 H_2O, separates when a solution of neptunium(VI) in dilute nitric acid is evaporated; this forms the hexahydrate on exposure to moist air. Hydrated neptunium trioxide dissolves in liquid nitrogen pentoxide and evaporation of the resulting solution yields[34] the adduct $NpO_2(NO_3)_2 \cdot N_2O_5 \cdot H_2O$. Hydrated plutonyl(VI) nitrate, $PuO_2(NO_3)_2 \cdot 6\ H_2O$, is obtained[55] by evaporating an aqueous nitric acid solution of plutonium(VI); small crystals of the compound are pink and large ones are brown. This hexahydrate loses water on standing over phosphorus pentoxide in a vacuum or when heated at 130°; as with uranyl(VI) nitrate, the di- and tri-hydrate is formed. In contrast to the behaviour of the uranyl(VI) compound, prolonged heating at $150°$ is reported to yield anhydrous plutonyl(VI) nitrate[55].

Uranyl nitrate hexahydrate possesses orthorhombic symmetry (space group $Cmc\,2_1$) and the compound has been shown by X-ray[56] and neutron[57] diffraction to have the constitution $[UO_2(NO_3)_2(H_2O)_2]\cdot 4\,H_2O$. The structure consists of an approximately hexagonal bipyramidal arrangement of eight oxygen atoms around the uranium atom, the two bidentate nitrate groups and the two ҫoordinated water molecules being in *trans* positions in the equatorial plane, with the uranyl, O=U=O, group normal to this plane[56,57](Fig. 19). The corresponding dihydrate has the same structural arrangement[58].

Anionic trinitratocomplexes, $M^I[M^{VI}O_2(NO_3)_3]$, $[M^I = (C_2H_5)_4N$, Cs, Rb, *etc.*; M^{VI} = U (yellow), Np (dark-purple to brown), Pu (brown)] are quite well known. They are prepared from solutions of the actinide(VI) in concentrated nitric acid on addition of the appropriate M^INO_3 (*e.g.*, tetraethylammonium salts[40]). A neutron diffraction study of $RbUO_2(NO_3)_3$, which possesses rhombohedral symmetry (space group $R\bar{3}c$), has shown that the uranium atom is surrounded by eight oxygen atoms at the corners of a distorted hexagonal bipyramid[59], as in the structures of the uranyl nitrate hydrates mentioned above. The caesium compound has the same structure as the rubidium one[59a].

Tetranitratocomplexes, $M_2^IUO_2(NO_3)_4$ (M = $(C_2H_5)_4N$[60], NH_4, Rb, Cs^{202}) are also known. They are obtained by fusing together the stoicheiometric quantities of the components $[(C_2H_5)_4N]$ or by evaporating aqueous solutions of them. The alkali metal salts are of monoclinic symmetry (space group $P2_1/c$) and the anion is a hexagonal bipyramid, with two *trans* bidentate and two *trans* unidentate nitrate groups in the equatorial plane[202].

Monodentate oxygen or nitrogen donor ligands form *bis* complexes with uranyl nitrate; the best known complexes are listed in Table 24. They are usually prepared from the ligand and uranyl nitrate hexahydrate in a non-aqueous solvent such as acetone, methyl cyanide, ethanol or diethyl ether, but the DMF complex (Table 24) is obtained from aqueous solution. The alkyl phosphate complexes are conveniently

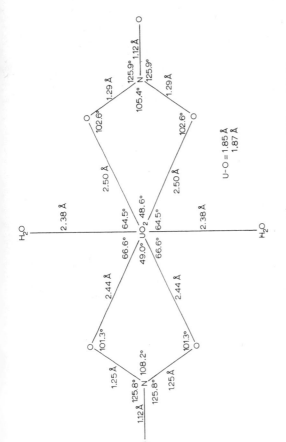

Fig. 19. The coordination about the uranyl, UO_2^{2+}, group in $UO_2(NO_3)_2 \cdot 6\,H_2O$ (D.A. Hall, A.D. Rae and T.N. Waters, *Acta Cryst.*, 19 (1965) 389). Reproduced by permission of the International Union of Crystallography.

TABLE 24

SOME URANYL NITRATE COMPLEXES, $UO_2(NO_3)_2 \cdot 2L$

Ligands		
Alkyl or aryl phosphates	$(RO)_3PO$	R = Et[61], nBu[62], iBu, secBu, iamyl, 3amyl[63], Ph[64]
Phosphine oxides	R_3PO	R = Ph[61],Me$_2$N[65]
Arsine oxides	R_3AsO	R = Ph[61]
Sulphoxides	R_2SO	R = Me[42],Ph[64, 66], thioxane[67]
Miscellaneous		Pyridine[68, 69], N,N-dimethylformamide(DMF)[70]
		quinoline N-oxide[71].

Abbreviations; Me, CH_3; Et, C_2H_5; Bu, C_4H_9; Ph, C_6H_5.

prepared by solvent extraction; for example, a benzene solution of the ligand is shaken with a dilute nitric acid solution of uranyl nitrate[61], and evaporation of the benzene phase yields the complex. Alternatively, the pure ligand is shaken with solid uranyl nitrate hexahydrate until a saturated solution is obtained[62, 63]. This saturated solution is actually the liquid *bis* complex. In the case of tri-*n*-butyl phosphate, the complex $UO_2(NO_3)_2 \cdot 2$ TBP prepared by this last method melts sharply[62] at $-6 \pm 0.5°$. The *bis* TBP complexes are important because they are the actinyl(VI) species extracted from nitric acid media in the chemical processing of neutron-irradiated uranium.

A few complexes with bidentate ligands are also known; the orthophenanthroline (phen) complex, $UO_2(NO_3)_2 \cdot (phen)$, is prepared from ethanolic solutions of the components[61], but phenacyldiphenylphosphine oxide, $Ph_2P(O)CH_2C(O)Ph$, forms a *bis* complex in which the ligand is unidentate[72]. Hydrated pyridine and other heterocyclic amine[73], 2,2-dipyridyl and orthophenanthroline[74] complexes are also known; the latter are hemihydrates, $UO_2(NO_3)_2 \cdot L \cdot 0.5 H_2O$, precipitated when an ethanolic solution of uranyl nitrate hexahydrate is added to an ethanol solution of the ligand[74]. The tetrahydrated and anhydrous dipyridyl complexes of uranyl(VI) nitrate have also been recorded[75].

The uranyl nitrate–urea system is more complicated than the systems with other ligands; the complexes formed in this system range

from $[UO_2(NO_3)_2(urea)_2]$ to $[UO_2(urea)_5](NO_3)_2$. They are prepared
from ethanol solutions of the components and by grinding together the
stoicheiometric quantities of urea and uranyl nitrate hexahydrate[76]. In
one of these complexes, $[UO_2(H_2O)(urea)_4](NO_3)_2$, the cation has been
shown[58] to be an almost regular pentagonal bipyramid. Presumably the
pentakis urea complex will have a similar structure. However, the uranium
atom is 8-coordinate in the *bis* triethylphosphate complex, $UO_2(NO_3)_2$-
$[(EtO)_3PO]_2$, which is of triclinic symmetry[77], and in the *bis*
triphenylphosphine and arsine oxide complexes, which are of monoclinic
symmetry[78] (space group $P2_1/c$). The arrangement is approximately
hexagonal bipyramidal, with the oxygen-donor ligands *trans*, as in the
structures of the hydrates.

6. NITRITES

The only recorded nitrite is the thorium(IV) complex with
N,N-dimethylformamide (DMF), $Th(NO_2)_4 \cdot 4$ DMF. This is a slightly
yellowish solid, which crystallises when a DMF solution of hydrated
thorium tetranitrate is saturated with sodium nitrite[79]. Some DMF is
lost at $224°$ and the product melts at $265°$.

7. PHOSPHATES

(a) Actinide(III) compounds

Phosphates of the tervalent actinides are known for actinium,
plutonium, americium and curium. White $AcPO_4 \cdot 0.5 \ H_2O$ is
precipitated from actinium(III) solutions by [79a] sodium dihydrogen
phosphate and pale blue $PuPO_4 \cdot 0.5 \ H_2O$ precipitates[80] when aqueous
solutions containing tervalent plutonium and diammonium phosphate,
$(NH_4)_2HPO_4$, are respectively added simultaneously to dilute (0.07 M)
hydrochloric acid at $80-90°$. The corresponding dark-rose americium
and white curium compounds[81] are precipitated from dilute acid

solution by disodium or diammonium phosphate at room temperature. The actinium, plutonium and americium compounds are isomorphous with lanthanum phosphate hemihydrate, $LaPO_4 \cdot 0.5\ H_2O$ (hexagonal, space group $C6_2 2-D_4^6$) and lose water at $950°$ and $200°$ respectively to yield the anhydrous compounds; these are of monoclinic symmetry. The anhydrous americium(III) compound can also be prepared by heating together the stoicheiometric quantities of americium dioxide and diammonium phosphate at[81] $600-1000°$.

(b) Actinide(IV) compounds

The addition of orthophosphoric acid, H_3PO_4, to aqueous thorium(IV) nitrate yields a white gelatinous precipitate which, when dried at $105°$, has the composition $Th_3(PO_4)_4 \cdot 12\ H_2O$. This can be dehydrated completely at $850°$ in a vacuum [82]. The green uranium analogue is precipitated from uranium(IV) chloride solutions by aqueous trisodium phosphate; it dehydrates when heated in hydrogen[83]. With an excess of phosphoric acid or alkali phosphate present, the acid phosphate, $Th(HPO_4)_2 \cdot 1-2\ H_2O$ and $Th(HPO_4)(H_2PO_4)_2 \cdot 2\ H_2O$, are formed[84]. The gelatinous green uranium(IV)[85], pale-green neptunium(IV)[86] and white plutonium(IV)[87] phosphates, $M(HPO_4)_2 \cdot x\ H_2O$, are precipitated when sodium dihydrogen phosphate (U) or phosphoric acid (Np, Pu) is added to an aqueous solution of the quadrivalent metal. When the plutonium compound is heated at $110°$ in $1.8\ M\ HNO_3-0.4\ M\ H_3PO_4$ in a sealed tube, hydrated $Pu_3(PO_4)_4$ results, whereas in $1.45\ M\ HNO_3-1.0\ M\ H_3PO_4$ at $110°$ the product is $Pu_2H(PO_4)_3 \cdot x\ H_2O$, a red solid. It has also been reported[88] that the addition of solutions of plutonium(IV) in orthophosphoric acid to ethanol yields pink, gelatinous precipitates of composition $H_6Pu(HPO_4)_5 \cdot x\ H_2O$ and $H_4Pu(HPO_4)_4 \cdot x\ H_2O$, depending on the conditions. The precise nature of these compounds is unknown.

Alkali metal orthophosphatocomplexes of the form $M^IM^{IV}(PO_4)_3$ (M^{IV} = Th, U) are prepared by heating a homogeneous mixture of the alkali metal-phosphate and the actinide metal dioxide at $1200°$, using B_2O_3 as a flux. The product is purified by washing the cooled melt with boiling water, which does not affect the phosphatocomplexes[89]. The sodium salt[90], $NaTh_2(PO_4)_3$, is isostructural with its potassium[91] analogue. Both are of monoclinic symmetry (space group Cc) and the thorium atom is 9-coordinate; the geometry is a very badly distorted tricapped trigonal prism. This bears some resemblance to a pentagonal bipyramid in which there are two, instead of one, oxygen atoms on each side of the pentagon[91].

The actinide(IV) metaphosphates, $M(PO_3)_4$, are prepared by fusing a tetrahalide (Th[92]; white) or dioxide (U[83]; green) with metaphosphoric acid or by heating[93] a uranium oxide (UO_2, U_3O_8) with phosphorus pentoxide at 400-1100°. The uranium compound is also obtained[93] by treating UP, UO_2 or U_3O_8 with concentrated orthophosphoric acid and heating the liquid so produced at 400°. In each case the solid obtained is leached with water to remove excess phosphorus pentoxide or metaphosphoric acid. Crystals of the uranium and plutonium (pink to greenish-tan) compounds can be grown from solutions of the dioxide in metaphosphoric acid; they are of orthorhombic symmetry[94] (space group $Pnab$-D_{2h}^{14}). A transition to a second crystal modification occurs[95] at 750° (Th) or 810° (U).

White, hydrated thorium pyrophosphate, $ThP_2O_7 \cdot x\, H_2O$, is precipitated when aqueous sodium pyrophosphate, $Na_4P_2O_7$, is added to an aqueous solution of quadrivalent thorium; it is completely dehydrated[96] at 540°. The hydrated uranium(IV) compound is obtained in the same way as the hydrated thorium one [105], but the green anhydrous compound, UP_2O_7, is best prepared by heating[83,93] the metaphosphate at about 1200°. The neptunium analogue (αThP_2O_7 structure) results from heating to 600° the dark green precipitate formed when disodium phosphate is added to a solution containing tervalent neptunium[81], whereas the plutonium compound is prepared by heating the mixed plutonium(IV) oxalatophosphates in air[80]. These oxalatophosphates are obtained as pink precipitates, for example by adding a solution of plutonium(IV) oxalate in 95% orthophosphoric acid to 5 M oxalic acid solution at 70-100°; they are of variable composition and may not be definite compounds. Plutonium pyrophosphate is stable to 1000° in air but decomposes slowly in a vacuum at this temperature, yielding $PuPO_4$.

Thorium(IV) and uranium(IV) pyrophosphates are dimorphic[97]; the low temperature (α) form is of cubic symmetry and the high temperature (β) form, which possesses orthorhombic symmetry, is stable only between $810°$ and $1290°$. At higher temperatures still reversion to the α-form occurs but decomposition begins above $1300°$.

The anionic complex[12], $Na_4U(P_2O_7)_2 \cdot 8 H_2O$, an apple green solid, is precipitated when cold, saturated aqueous sodium pyrophosphate is added dropwise to a saturated aqueous solution of uranium(IV) sulphate at 50-$60°$.

(c) Actinide(V) compounds

The only recorded compounds are the white protactinium phosphate[114], $PaO(OH)HPO_4 \cdot x H_2O$ ($x = 1$-3) and the white ammonium plutonium salt, $NH_4PuO_2HPO_4 \cdot 4 H_2O$. The former is precipitated when $0.14 M H_3PO_4$ is added to a solution of protactinium(V) in $2.5 N H_2SO_4$ and the latter when diammonium phosphate is added to an aqueous solution of plutonium(V) in very dilute nitric acid[17].

(d) Actinide(VI) compounds

The yellow neutral uranyl phosphate, $(UO_2)_3(PO_4)_2 \cdot 4 H_2O$, is precipitated when orthophosphoric acid is added to aqueous uranyl acetate provided that an excess of the latter is present[98]; it is also formed[99] by digesting $NH_4UO_2PO_4 \cdot 3 H_2O$ in $0.8 M$ nitric acid at $100°$. The bright-yellow anhydrous compound was originally reported[98] as being obtained by heating the tetrahydrate to $130°$, but later work[100] indicates that the monohydrate is formed at this temperature and that the anhydrous compound, which is stable to $500°$, is obtained above $250°$.

The bright-yellow acid phosphate, $UO_2HPO_4 \cdot 4 H_2O$, is prepared in the same way as the neutral phosphate but with a slight excess of orthophosphoric acid[98,99]; it loses water to form the dihydrate[99,100] at $110°$ and the anhydrous compound is obtained[100] at $150°$. An alternative method of preparing the compound is by adding 89% orthophosphoric acid dropwise to a boiling aqueous solution of uranyl acetate[101]. The UO_2^{2+}-H_3PO_4 system has been studied in some detail[102] and the concentration ranges of the reactants for the precipitation of these two phosphates have been established. A second acid phosphate, $UO_2(H_2PO_4)_2 \cdot 3\ H_2O$, is obtained[99] by adding the monohydrogen compound to 85% orthophosphoric acid.

The monohydrogen phosphate forms a calcium salt, $Ca(UO_2)_2(PO_4)_2 \cdot x\ H_2O$, identical with the mineral autunite[101], and a number of insoluble alkali metal salts, such as $NH_4UO_2(PO_4) \cdot 3\ H_2O$ and $Na_4UO_2(PO_4)_2$, are also known[98].

The corresponding plutonium(VI) phosphate system is less well known; $PuO_2(HPO_4) \cdot 4\ H_2O$, a pale-green, hygroscopic solid, is precipitated from mixtures of 0.085 M orthophosphoric acid and plutonium(VI) in 0.1 M nitric acid. It is appreciably soluble in this medium and the precipitation yield is improved by adding methanol[103]. The solubility of the compound in aqueous orthophosphoric acid increases with the acid concentration, and an orange-red solution results in 5.88 M H_3PO_4. When this is poured into absolute methanol, a pink gelatinous precipitate separates[104] which, when dried, has the composition $PuO_2(H_2PO_4)_2 \cdot x\ H_2O$. An ammonium salt[103], $NH_4PuO_2PO_4 \cdot 3\ H_2O$, separates as a green precipitate when aqueous diammonium phosphate is added to a hot 0.1 M nitric acid solution containing plutonium(VI). This loses water on standing over phosphorus pentoxide, yielding the anhydrous salt at 180°, which decomposes to PuO_2HPO_4 at 410°.

Hydrated uranyl(VI) pyrophosphate, $(UO_2)_2P_2O_7 \cdot 4\ H_2O$, is precipitated from uranyl(VI) nitrate solution on the addition of aqueous sodium pyrophosphate[105]. The anhydrous uranyl and plutonyl(VI) compounds result when the phosphates MO_2HPO_4 are heated at 600° (U^{100}) and 625° (Pu^{103}).

8. PHOSPHITES AND HYPOPHOSPHITES

According to Flahaut[106], thorium phosphite, $Th(HPO_3)_2 \cdot 8\ H_2O$, hypophosphite, $Th(H_2PO_2)_4 \cdot H_2O$, and hypophosphate, ThP_2O_6, are precipitated by mixing the appropriate sodium salt with aqueous thorium(IV) nitrate solution. Uranyl(VI) phosphite, $UO_2HPO_3 \cdot 2 H_2O$, is apparently precipitated as a yellow jelly when phosphorous acid, H_3PO_3, is added to aqueous uranyl(VI) acetate, and this slowly becomes crystalline on standing[107]. Precipitation also occurs with disodium phosphite, but oxidation to uranyl(VI) phosphate is reported to occur[108], and the system requires further investigation. The corresponding reaction of sodium hypophosphite with aqueous uranyl(VI) nitrate is reported[109] to yield a precipitate of $UO_2(H_2PO_2)_2 \cdot 3 H_2O$ or, when an excess of the sodium salt is used, a precipitate of $Na(UO_2)_2$-$(H_2PO_2)_5 \cdot x H_2O$. The potassium analogue of the latter is apparently precipitated as the anhydrous salt[109].

9. ARSENATES

Light-green plutonium(III) and pale-rose americium(III) arsenates, $MAsO_4$, are both isostructural with the lanthanide arsenates[81]. The former is obtained as a precipitate when diammonium arsenate is added to a 0.1 M hydrochloric acid solution of plutonium(III) and the latter when an equimolar mixture of americium(III) nitrate and diammonium arsenate is heated at 1000°.

Thorium(IV) arsenate, $Th(HAsO_4)_2 \cdot x H_2O$ ($x = 5$[110] or 6[110,111]), is precipitated from thorium(IV) nitrate solution by aqueous arsenic acid, H_3AsO_4. With an excess of the acid present a tetrahydrate is also formed[111], but this readily converts to the hexahydrate in water. The monohydrate[111a] is obtained when a solution of $Th(NO_3)_4 \cdot 5 H_2O$ in 3.6 M nitric acid is mixed with 5 M arsenic acid and the mixture is heated under reflux for 3–4 days. The anhydrous compound results when the monohydrate is heated to 240°; this appears to yield the pyroarsenate, $ThAs_2O_7$, at 425°. The monohydrate has ion-exchange properties and is specific for lithium[111a] which presumably forms a lithium arsenatothorate. The uranium(IV) arsenate, $U(HAsO_4)_2 \cdot 2 H_2O$, a pale green solid, is precipitated in the same way as the thorium compound from a uranium(IV) solution in dilute sulphuric acid, but is apparently best prepared by treating freshly precipitated uranium(IV) hydroxide with dilute aqueous arsenic acid[112].

Uranyl(VI) arsenate, $UO_2HAsO_4 \cdot x\ H_2O$, where x is 3[113] or 4[101, 113], is precipitated from a cold solution of uranyl(VI) acetate on addition of the stoicheiometric quantity of arsenic acid. The degree of hydration appears to depend on the quantities of the reagents which are used; the compound dehydrates at $120°$. With an excess of uranyl(VI) acetate, the product is $(UO_2)_3(AsO_4)_2 \cdot 12\ H_2O$, identical to the mineral trögerite[113]. The corresponding precipitation reaction with alkali metal arsenates yields the salts $M^IUO_2AsO_4(M^I = Na, K, NH_4)$; calcium and copper(II) arsenates yield[113] the salts $M^{II}(UO_2AsO_4)_2$. The calcium salt is found in nature as the mineral uranospinite[101].

10. SULPHATES

(a) Actinide(III) compounds

The hydrated actinide(III) sulphates, $M_2(SO_4)_3 \cdot x\ H_2O$ (U, $x = 8$, olive green[115]; Pu, $x = 7$ (?), violet[116]; Am, $x = 5$, pink[118]) are precipitated when ethanol is added to a solution of the tervalent actinide in dilute sulphuric acid. The blue pentahydrated plutonium compound is obtained by the spontaneous evaporation of a solution of plutonium(III) in $2\ M\ H_2SO_4$ and is isomorphous with $Nd_2(SO_4)_3 \cdot 5\ H_2O$, which possesses monoclinic symmetry[117]. The light blue-grey anhydrous plutonium[116] and white americium[119] compounds are obtained by heating the hydrates to $130-150°$ and to $550-650°$ respectively; the americium compound decomposes[119] above $730°$ forming AmO_2. The uranium(III) compound is very readily oxidised in air and in aqueous solution oxidation to uranium(IV) is also rapid[115].

Hydrated sulphatocomplexes of composition $M^IM^{III}(SO_4)_2 \cdot x\ H_2O$ are known for actinium[79a] (white, $M^I = K$), uranium[120] (olive-green, $M^I = NH_4$, Na, Rb, Cs), neptunium[203] (blue-violet), plutonium[121] (light-blue or lavender, $M^I = NH_4$, K, Rb, Cs, Tl), americium[118] (rose, $M^I = K$, Rb, Cs, Tl) and curium[122] (white, $M^I = K$). They are prepared by adding the alkali sulphate to a solution of the actinide(III) in dilute sulphuric or hydrochloric acid, precipitation being accelerated by adding ethanol to the mixture (Pu, Am) or by saturating the solution with alkali sulphate (Np, Cm). By use of the appropriate quantities of alkali sulphate complexes of composition $K_3M^{III}(SO_4)_3 \cdot x\,H_2O$ (U^{120}, Am^{118}), $K_5M^{III}(SO_4)_4$ (U, tobacco-coloured[123]; Np, blue-violet[203]; Pu, off-white[121]) and $M_8^I\,M_2^{III}(SO_4)_7$ (Pu, M^I = K, pale-lavender[117]; Am, M^I = K, Cs, Tl, rose[118]) can also be obtained. No structural data are available for any of these compounds.

(b) Actinide(IV) compounds

$Th(SO_4)_2 \cdot 9\,H_2O$ appears to crystallise from aqueous thorium(IV) solutions in dilute sulphuric acid whereas crystallisation from neutral solution at 20-25° yields the octahydrate. Lower hydrates are obtained[124] by drying these products at 100-110° and the anhydrous compound[125] is formed below 400°. The solubility of the thorium compound in sulphuric acid decreases with increasing acid concentration. The uranium(IV) sulphate hydrates appear to be similar to the thorium compounds; green $U(SO_4)_2 \cdot 9\,H_2O$ is said to crystallise from aqueous solution, but it apparently loses water to form the octahydrate on exposure to air[126]. Later work indicates that the latter is obtained from dilute acid solution and that the tetrahydrate crystallises from the concentrated acid[127] or when the acid solution is heated at 70°. The tetrahydrate loses water in three stages, forming[128] the trihydrate at 125°, the hemihydrate at 165° and the anhydrous compound at 220°. Decomposition to uranyl sulphate occurs above 450°. Bright green hydrated neptunium(IV) sulphate crystallises when a solution of neptunium(IV) in 0.5 M H_2SO_4 is evaporated; the degree of hydration is uncertain[129]. The light coral-pink to light red-brown plutonium(IV) compound, $Pu(SO_4)_2 \cdot 4\,H_2O$, separates from solutions of Pu(IV) in 1.7 M H_2SO_4 on the addition of methanol. It yields[130] a grey lower hydrate, $Pu(SO_4)_2 \cdot 0.3\,H_2O$, at 150°.

Crystals of $U(SO_4)_2 \cdot 4\,H_2O$ possess orthorhombic symmetry (space group *Pnma*) and the uranium atom is 8-coordinate, being surrounded by the four oxygen atoms of the water molecules, and by four more from four sulphate groups, at the corners of a square antiprism[131].

Basic salts of composition $M(OH)_2SO_4$ are known for thorium[132] and uranium[133]; they are made by heating aqueous solutions of the disulphate in a sealed tube at 105–115° (Th) or 100–150° (U). The uranium compound is isomorphous with the thorium one, which possesses orthorhombic symmetry (space group $Pnma$-D_{2h}^{16}); the structure of the latter consists of zigzag chains of $Th(OH)_2$ groups held together by the sulphate groups. The thorium atom is 8-coordinate, made up of four atoms from hydroxyl groups and four from sulphate groups; the coordination geometry is a square antiprism[134]. A light grey-green basic plutonium(IV) sulphate, $Pu_2O(SO_4)_3\cdot8\ H_2O$, separates from plutonium(IV) sulphate solutions in $0.07\ M\ H_2SO_4$ in the presence of sodium sulphate and methanol[130]; no structural information is available for this compound.

The solubilities of the disulphates in aqueous media increase in the presence of alkali sulphates because of the formation of anionic sulphatocomplexes. Thorium sulphatocomplexes of composition $M_6^I Th(SO_4)_5\cdot3\ H_2O$ (M^I = NH_4, Cs) separate from the $M_2^I SO_4$–$Th(SO_4)_2$–H_2O system when the concentration of $M_2^I SO_4$ is high and $(NH_4)_8$-$Th(SO_4)_6\cdot2\ H_2O$ crystallises when solid $Th(SO_4)_2\cdot8\ H_2O$ is added to a saturated solution of ammonium sulphate at 50–60°. Thermogravimetric data for these salts are available and it has been shown that dehydration occurs at moderate temperatures. For example[135] $Cs_6 Th(SO_4)_5\cdot3\ H_2O$ becomes anhydrous above 180°. Analogous plutonium compounds[136], $M_6^I Pu(SO_4)_5\cdot x\ H_2O$ (M^I = Na, NH_4) separate on standing when $Pu(SO_4)_2$ is dissolved in $1\ N\ H_2SO_4$ and the solution is saturated with the alkali metal sulphate. The sodium compound is yellow-rose and the ammonium one is brick-red. The cobalt(III) hexammine salts, $[Co(NH_3)_6]_2[M(SO_4)_5]\cdot x\ H_2O$, of the thorium(IV) and plutonium(IV) complexes are orange and greenish-yellow respectively. They are precipitated when a solution of a cobalt(III) hexammine salt is added to a solution of the actinide(IV) in aqueous ammonium sulphate[140].

The more usual sulphatocomplex species isolated from aqueous solution are of the composition $M_4^I M^{IV}(SO_4)_4\cdot x\ H_2O$. Crystals of

$K_4Th(SO_4)_4 \cdot 2\,H_2O$ are known to possess triclinic symmetry[137] (space group $C\bar{1}$), but it is not certain whether the analogous grey-green plutonium compound is isomorphous with the thorium one. The corresponding rubidium and caesium plutonium(IV) tetrasulphatocomplexes apparently separate from aqueous media as, respectively, the lilac-rose and pale-rose anhydrous compounds. However, both form green hydrates in air[136]. The green hydrates are also reported to be formed when an alkali sulphate ($M = NH_4$, K) solution is added to a methanolic solution of plutonium(IV) sulphate or when $Pu(SO_4)_2 \cdot 4\,H_2O$ is dissolved in nearly neutral rubidium sulphate solution and ethanol is added to the mixture[139]. However, $Pu(SO_4)_4 \cdot 4\,H_2O$ precipitates when the concentration of plutonium exceeds 50 g/l. The u.v. visible spectra of the plutonium compounds have been published[136] and infrared spectral data for the thorium complexes are also available.

A dimethyl sulphoxide complex of composition $Th(SO_4)_2 \cdot 4\,DMSO \cdot 3\,H_2O$ is reported[46] to be formed when $Th(SO_4)_2 \cdot 8\,H_2O$ is treated with the ligand. Urea complexes of thorium[199] and uranium[200] disulphates, and an acetamide complex of the latter[200], are also known.

(c) Actinide(V) compounds

The only actinide(V) sulphates known are the basic protactinium compounds[142] $H_3PaO(SO_4)_3$ and $HPaO(SO_4)_2$. The white trisulphatocomplex crystallises when a solution of protactinium(V) in a mixture of hydrofluoric and sulphuric acids is evaporated until the hydrofluoric acid is completely eliminated. This compound decomposes to the white disulphatocomplex at $375°$ and there is thermogravimetric evidence for the formation of $HPaO_2SO_4$ at about $500°$; decomposition to protactinium pentoxide occurs at $600°$.

(d) Actinide(VI) compounds

A variety of hydrates of uranyl sulphate have been reported; the trihydrate, $UO_2SO_4 \cdot 3\,H_2O$, is obtained by evaporating a saturated solution of uranyl nitrate with concentrated sulphuric acid to dryness, redissolving the residue in water, and leaving the solution to crystallise. This hydrate loses water in three stages when heated, yielding the anhydrous compound above $224°$ according to one author[128] or above $300°$ according to another[143]. The phase relationships in aqueous uranyl sulphate solutions have been reported[144] and although these results favour the monohydrate and trihydrate as the stable phase, it seems[145] that the trihydrate is really $UO_2SO_4 \cdot 2.5\ H_2O$. Further investigation of this hydrate system is necessary to resolve the problem.

Sulphatocomplexes of the type $M_2^I UO_2(SO_4)_2 \cdot 2\ H_2O$ (M^I = K, NH_4, Rb) and $K_4UO_2(SO_4)_3 \cdot 2\ H_2O$ were reported many years ago, prepared by dissolving potassium uranate in sulphuric acid, or by mixing solutions of the components in the correct proportions, and evaporating the solutions in a vacuum[146]. Orange cobalt hexammine salts of the trisulphato-actinyl(VI) ion, $[Co(NH_3)_6HSO_4]_2\ MO_2(SO_4)_3 \cdot 5\ H_2O$ ($M = U^{140}$, Np^{147}, Pu^{140}, Am^{147}), are precipitated when a solution of a cobalt hexammine salt is added to a solution of the actinide(VI) in aqueous ammonium sulphate. The four compounds are isostructural. Compounds of the type $M_2^I (UO_2)_2(SO_4)_3 \cdot x\ H_2O$ crystallise from aqueous solutions of the components. The yellow ammonium salt ($x = 5$) is of orthorhombic symmetry[148] (space group probably $Pnma-D_{2h}^{16}$). The analogous plutonium compound is also a pentahydrate and is isolated as a yellow solid from sulphuric acid solutions of the corresponding trinitratocomplex[117]. The anhydrous caesium uranyl(VI) salt[149] is of tetragonal symmetry (space group $P\bar{4}2_1m-D_{2d}^3$). In this compound, sheets of $[(UO_2)_2(SO_4)_3]$ groups make up a layer structure held together by Cs^+ ions. The uranium atom is 7-coordinate in a pentagonal bipyramidal arrangement in which the UO_2^{2+} group is coordinated to 5 sulphate oxygen atoms in the equatorial plane (Fig. 20). The compound is prepared by slow evaporation of an aqueous solution of the components.

Oxygen-donor complexes of uranyl sulphate of composition from $UO_2SO_4 \cdot L$ to $UO_2SO_4 \cdot L$ (L = acetamide, urea, etc.) and complexes with the bidentate nitrogen-donors dipyridyl (dipy) and orthophenanthroline (phen), $UO_2SO_4 \cdot L \cdot x\ H_2O$ (x = 4 and 5 respectively) are also known[74].

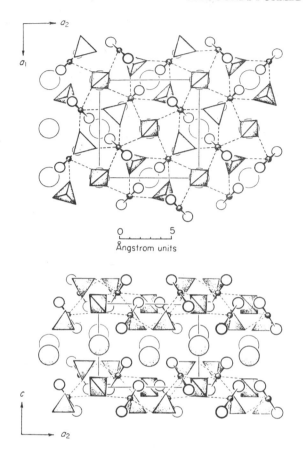

Fig. 20. Schematic views along [001] and [100] of the crystal structure of $Cs_2(UO_2)_2(SO_4)_3$. Uranyl groups are represented by the dumbbell-shaped figures, caesium atoms by large open circles and sulphate groups as tetrahedra. The coordination of sulphate groups to uranyl groups is indicated by broken lines (M. Ross and H.T. Evans Jr., *J. Inorg. Nucl. Chem.*, 15 (1960) 338). Reprinted by permission of Maxwell Scientific Int., Inc. Elmsford, N.Y.

Partially hydrolysed species of composition $(UO_2)_2(OH)_2SO_4 \cdot L_2 \cdot 2\,H_2O$ (L = dipy, phen, respectively green and yellow-green) are precipitated when the solid ligand is added to aqueous uranyl sulphate solution[150]. The nature of these basic compounds is unknown.

11. SULPHITES

(a) Actinide(III) compounds

A hydrated plutonium(III) sulphite, $Pu_2(SO_3)_3 \cdot x\ H_2O$, is obtained as a blue precipitate when aqueous ammonium sulphite is added to a solution of tervalent plutonium in 1 M hydrochloric acid in the presence of sulphur dioxide[151]. No other actinide(III) sulphites have been reported.

(b) Actinide(IV) compounds

A white, hydrated thorium(IV) sulphite, $Th(SO_3)_2 \cdot 4\ H_2O$, is deposited[124] on heating the solution obtained by dissolving hydrated thorium dioxide in water saturated with sulphur dioxide. It is easily hydrolysed[152]. The green precipitate which results[151] when ammonium sulphite is added to a plutonium(IV) solution at pH 4 is presumably a sulphite, but both this and the thorium(IV) compound may be basic sulphites. Sulphitocomplexes such as $(NH_4)_4M^{IV}(SO_3)_4$ (M^{IV} = Th, U) and $M_6^I U_2(SO_3)_7 \cdot x\ H_2O$ (M^I = Na, K) separate on cooling the solution which results from heating the solid disulphate hydrate with a concentrated aqueous solution of the alkali metal sulphite[12]. Other salts of the form $M_2^I [Th(SO_3)_3(H_2O)_2] \cdot x\ H_2O$ (M^I = Na, K[152]), as well as the basic compounds of composition $M_4^I Th_2(OH)_2(SO_3)_5 \cdot x\ H_2O$ and $CN_3H_6[U(SO_3)_2OH] \cdot 2\ H_2O$ (CN_3H_6 = guanidinium[12]) have been recorded, but the preparative conditions are somewhat uncertain and the systems merit reinvestigation.

(c) Actinide(VI) compounds

Hydrated uranyl sulphite, $UO_2SO_3 \cdot 4.5\ H_2O$, is best prepared by saturating an aqueous suspension of uranium trioxide with sulphur dioxide at room temperature[153]. The trihydrate is obtained when the aqueous suspension is maintained at 80°. The compound is also obtained, with varying degrees of hydration, by the reaction of

References p. 177

aqueous solutions of uranyl salts with sulphur dioxide or alkali metal sulphite. The orange precipitate formed when ammonium sulphite is added to a solution of plutonium(VI) may be the corresponding plutonyl compound[151], although reduction to a lower oxidation state seems more probable. Mixed oxalatosulphitocomplexes, $M_2^I UO_2(SO_3)(C_2O_4)$ are apparently formed when solid uranyl oxalate, $UO_2C_2O_4 \cdot 3\ H_2O$, is mixed with the stoicheiometric quantity of solid alkali metal or guanidinium sulphite and a little water is added to the mixture[154]. The complexes $K_4UO_2(SO_3)_3$ and $K_2(UO_2)_2(SO_3)_3$ are also known.

12. THIOSULPHATES

Anhydrous thorium thiosulphate, $Th(S_2O_3)_2$, is precipitated when methanol is added to a solution containing a thorium(IV) salt and sodium thiosulphate[155]. It decomposes to thorium dioxide above $800°$. No other actinide thiosulphates are known.

13. SELENATES AND SELENITES

Thorium selenate, $Th(SeO_4)_2 \cdot 9\ H_2O$, is prepared in the same way as the sulphate (p. 164). It forms the monohydrate at $120°$ and is not very soluble in water[124]. The corresponding selenite, $Th(SeO_3)_2 \cdot 8\ H_2O$, is precipitated from aqueous thorium(IV) sulphate[156] or chloride[124] by adding, respectively, sodium selenite or selenous acid. Protactinium(V) selenate[142], $H_3PaO(SeO_4)_3$, is prepared in the same way as the sulphate (p. 166) but, unlike the latter, there does not appear to be any diselenato- or monoselenatocomplex formed in the thermal decomposition of the compound, which yields the pentoxide at about $550°$.

Uranyl selenate itself is unknown, and evaporation of aqueous uranyl nitrate with selenic acid in a vacuum yields complex species such as $H_2[UO_3(SeO_4)] \cdot 2\ H_2O$. Treatment of this product with potassium selenate yields[157] a salt, $KH[UO_3(SeO_4)] \cdot H_2O$. In contrast to this behaviour, the yellow selenite, $UO_2SeO_3 \cdot 2\ H_2O$, is precipitated from solutions of uranyl salts by sodium selenite or selenous acid. It loses water above 100° but the anhydrous compound is best prepared by heating uranium trioxide with selenium dioxide at 350–450°. This decomposes above 550°. A selenitocomplex, $Na_2UO_2(SeO_3)_2$, is formed when the selenite is treated with concentrated aqueous sodium selenite[158].

14. TELLURATES AND TELLURITES

Thorium(IV) tellurate, a white gelatinous product of composition $ThOTeO_4 \cdot 8 H_2O$, is precipitated when sodium tellurate is added to aqueous thorium(IV) sulphate[159]. Four molecules of water are lost above about 100°. A uranyl(VI) tellurate of composition $UO_2TeO_4 \cdot 2 Na_2TeO_4 \cdot 4 H_2$ is precipitated when sodium tellurate is added to aqueous uranyl nitrate[160]; both compounds are presumably derived from orthotelluric acid (e.g. $ThH_2TeO_6 \cdot 7 H_2O$). A blue-green uranium(IV) tellurite, $U(TeO_3)_2$, is apparently obtained[197] from hydrochloric acid solutions of H_2TeO_3 and UCl_4, and the yellow uranyl compound, UO_2TeO_3, is precipitated when H_2TeO_3 is added to aqueous uranyl acetate[198].

15. HALATES

(a) Chlorates and bromates

The standard method of preparing aqueous solutions of the thorium(IV) and uranyl(VI) compounds is by treating the actinide sulphate with the calculated quantity of the barium halate. Evaporation of the filtrate appears to yield thorium(IV) chlorate, but decomposition occurs when solutions of the corresponding bromate are evaporated[124]. The uranyl compounds behave in the same way; hydrated uranyl chlorate, a dull-yellow solid, can be isolated by evaporation, and it is soluble in ethanol, chloroform and diethyl ether, but it explodes when an attempt is made to recrystallise it from either of the first two solvents[161]. The corresponding bromate solution yields a syrupy mass on evaporation and this apparently evolves bromine in the process[162]. The chlorates are particularly dangerous to handle because of the risk of an explosion.

(b) Iodates

The light-brown precipitate[163] formed when aqueous potassium iodate is added to solutions containing plutonium(III) is presumably the triiodate, $Pu(IO_3)_3$.

Hydrated thorium iodate, $Th(IO_3)_4 \cdot x\, H_2O$, originally thought to have the composition $4Th(IO_3)_4 \cdot KIO_3 \cdot 18\, H_2O$, is precipitated when iodic acid or alkali iodate is added to aqueous solutions of thorium(IV). It hydrolyses[164] when washed with water and decomposes above 300°. The corresponding neptunium(IV) compound, a bronze coloured solid, is precipitated in the same way[86]. The composition[165] of the pink plutonium(IV) compound which is precipitated by potassium iodate under similar conditions appears to be close to $Pu(IO_3)_4$ when the precipitate is washed with 0.5 N HNO_3 to remove any of the potassium salt incorporated in it and dried at 150°. Berkelium(IV) iodate is coprecipitated with cerium(IV) iodate[166] but has not been isolated as a pure compound.

A plutonium(V) iodate is obtained[193] as a white precipitate from nitric acid solutions of plutonium(V), but neither its composition nor the exact preparative details have been reported.

Hydrated uranyl iodate is precipitated when iodic acid or sodium iodate is added to a solution of uranyl nitrate in nitric acid; the yellow product can be recrystallised from water. Addition of concentrated aqueous potassium iodate to a concentrated solution of uranyl iodate, however, yields[167] the yellow salt $KUO_2(IO_3)_3 \cdot 3\, H_2O$.

(c) Perchlorates

Hydrated thorium perchlorate, $Th(ClO_4)_4 \cdot 4\, H_2O$, is prepared by dissolving freshly precipitated hydrated thorium dioxide in perchloric acid and evaporating the resulting solution[168]. It can be recrystallised from 0.2 N HClO$_4$. The attempted preparation of the analogous uranium(IV) compound by metathesis of uranium tetrachloride with silver perchlorate in acetone, followed by precipitation of the uranium(IV) compound from the filtrate with isopentane and vacuum drying of the precipitate at 55°, yields a shock-sensitive, explosive, green solid which contains some uranyl perchlorate[169]. However, a hydrated phosphato-perchlorate, $U(H_2PO_4)_2(ClO_4)_2 \cdot x\ H_2O$, is obtained by shaking solid $U(HPO_4)_2 \cdot x\ H_2O$ with $10\ M$ perchloric acid[170].

Hydrated thorium tetraperchlorate decomposes to a basic compound of composition[191] $ThO(ClO_4)_2 \cdot x\ H_2O$ at 275–285° and aqueous solutions of the latter contain the tetrameric $Th_4(OH)_6^{10+}$ ion[192].

Hydrated uranyl perchlorate, $UO_2(ClO_4)_2 \cdot x\ H_2O$ ($x = 5$[187] or 6[177]) is prepared by treating uranyl nitrate hexahydrate with perchloric acid, evaporating and then recrystallising the product from water. The analogous plutonium(VI) compound, $PuO_2(ClO_4)_2 \cdot 6\ H_2O$, is apparently prepared in a similar manner to plutonyl(VI) nitrate. Its visible spectrum in organic solvents (N,N-dimethyl formamide, dimethyl sulphoxide and ethanol) has been published[194].

Oxygen-donor complexes of thorium(IV) and uranium(IV) perchlorates appear to be reasonably stable to heat and do not seem to be shock-sensitive. The known adducts of the actinide(IV) and uranyl(VI) perchlorates are listed in Table 25. The antipyrine complexes are prepared from aqueous solution, either by treating a solution of the appropriate actinide nitrate, in the presence of the ligand, with ammonium perchlorate[171] or by treating an aqueous solution of the actinide perchlorate with the ligand[172], which amounts to the same reaction. In each case the complex crystallises from the solution on standing. The DMA and NMA complexes are prepared by metathesis of the corresponding chloride complex with silver perchlorate in acetone; the impure complex, precipitated from the filtrate by isopentane, is recrystallised from nitromethane and methylisobutylketone[41, 169]. The thorium(IV) and uranium(IV) complexes deflagrate violently above about 160°. The remaining complexes are prepared from the hydrated perchlorate and the ligand,

TABLE 25

OXYGEN-DONOR COMPLEXES OF THE ACTINIDE PERCHLORATES

Ligand	Complex	m.p., °C	Reference
Antipyrine, $COC_{10}H_{12}N_2$	$[ThL_6](ClO_4)_4$	272–5(d)[a]	171
	$[ThL_7](ClO_4)_4$	d.[a] 300	172
	$[UO_2L_5](ClO_4)_2$	194–6	171, 172
N,N-dimethylacetamide,	$[ThL_6](ClO_4)_4$	133–4(d)[a]	41
DMA	$[UL_6](ClO_4)_4$	132	169
N-methylacetamide, NMA	$[UL_6](ClO_4)_4$	146–154	169
	$[UL_8](ClO_4)_4$	115–20	169
Dimethylsulphoxide, DMSO	$[ThL_{12}](ClO_4)_4$	–	173
	$[ThL_6](ClO_4)_4$	exp.[a] 290	173
	$[UO_2L_5](ClO_4)_2$	exp.[a] 330	67, 174
	$[UO_2L_4](ClO_4)_2$	–	174
Dimethylselenoxide, DMSeO	$[UO_2L_5](ClO_4)_2$	–	175
Diphenylsulphoxide, DPSO	$[ThL_6](ClO_4)_4$	d.[a] 300	195
	$[UO_2L_5](ClO_4)_2$	–	64
	$[UO_2L_4](ClO_4)_2$	–	66
Thioxane oxide, TSO	$[UO_2L_5](ClO_4)_2$	–	67
Hexamethylphosphoric triamide, HMPA	$[UO_2L_5](ClO_4)_2$	–	65
Octamethylpyrophosphoramide,	$[ThL_4](ClO_4)_4$	–	176
OMPA	$[UO_2L_3](ClO_4)_2$	–	176
Pyridine N-oxide, pyNO	$[ThL_8](ClO_4)_4$	exp.[a] 350	177
	$[UO_2L_5](ClO_4)_2$	exp.[a] 319	177

[a]d., decomposes; exp., explodes.

either alone (DMSO[173, 174], OMPA[176]) or in acetone (DMSeO[175], DPSO[66, 195]), methyl cyanide (DPSO[64], HMPA[65]) or ethanol (pyNO[177]).

It has been suggested[173] that whereas the ligand in $[Th(DMSO)_{12}]$-$(ClO_4)_4$ is coordinated to the metal by way of the oxygen atom (some free ligand is present in the lattice), the ligand in $[Th(DMSO)_6(ClO_4)_4]$ is sulphur-bonded. A similar suggestion has been made[174] concerning the uranyl perchlorate complexes with DMSO, where the *pentakis* complex is thought to be oxygen-bonded and the *tetrakis* complex, obtained by heating the former to 170°, to be sulphur-bonded. This seems rather unlikely. All of these complexes of the metal perchlorates are dangerous when heated; care must be exercised in the preparation of the DMSO complexes from hydrated metal perchlorates for the reaction with the ligand is markedly exothermal. In view of these dangers, it is surprising that the tetraphenylborate ion has not been used extensively in place of the perchlorate ion when a non-complexing anion is desired. The only actinide tetraphenylborates that have been prepared are two complexes of the uranium(III) compound, in which the ligands are phenazone (Pz, p. 000) and 4-dimethylaminophenazone (DMPz). These are $U(Pz_6)(BPh_4)_3$ and $U(DMPz)_4(BPh_4)_3$, both of which are purple solids. They precipitate when a methanolic solution of $NH_4UCl_4 \cdot 5 H_2O$, together with the appropriate ligand, is mixed with sodium tetraphenylborate in methanol[196].

(d) Periodates

A white gelatinous precipitate of the thorium(IV) compound, $ThHIO_6 \cdot 5 H_2O$, forms slowly when a mixture of thorium nitrate and sodium periodate in nitric acid is heated on a water bath[178]. The yellow uranyl compound, $(UO_2)_2I_2O_9$, is said to be precipitated from uranyl nitrate solution on addition of aqueous potassium periodate[108] but with uranyl acetate and a saturated solution of potassium periodate the product is reported[179] to be $K_2(UO_2)_2I_2O_{10} \cdot 5 H_2O$. Nothing is known about these compounds.

16. PERXENATES

The orange americium(III) compound, $Am_4(XeO_6)_3 \cdot 40 H_2O$, is precipitated from aqueous solutions of the americium(III) carbonato-complex. It dissolves in acid with immediate oxidation to a mixture of americium(V) and (VI). Uncharacterised thorium(IV) and uranyl(VI) perxenates are also known[180].

References p. 177

17. COMPOUNDS OF *d*-TRANSITION METAL OXOACIDS

Most of these compounds can be obtained only by fusing together the appropriate oxides; a review discussing much of the recent work in this area has appeared[181] and the only compounds dealt with here are those which can be obtained from aqueous solution.

Thorium(IV) chromate, $Th(CrO_4)_2 \cdot H_2O$, an orange solid, separates when the solution obtained by dissolving freshly precipitated hydrated thorium dioxide in aqueous chromium trioxide is heated to $90°$, or when the solution is allowed to stand. The same product is obtained by boiling a solution containing thorium tetranitrate and potassium dichromate. The first stage in the precipitation is the formation of the golden-yellow basic chromate, $Th(OH)_2CrO_4 \cdot H_2O$, which is converted to the hydrated chromate in contact with aqueous chromium trioxide or potassium dichromate[182]. The basic chromate can be isolated by using very dilute potassium dichromate solution in the reaction, and the crystals are of monoclinic symmetry (space group $P2_1/m\text{-}C_{2h}^2$). The structure consists of infinite zigzag chains of $[Th(OH)_2]^{2+}$ groups held together by CrO_4^{2-} ions and water molecules, the thorium atom being 8-coordinate as in the basic sulphate[183] (p. 165). The yellow hydrated uranyl chromate, $UO_2CrO_4 \cdot 5.5\,H_2O$, is prepared by the action of warm, concentrated aqueous chromium trioxide on uranyl hydroxide[184]. The complex salts $M_2^{I}(UO_2)_2(CrO_4)_3 \cdot x\,H_2O$ (M^{I} = Na^{146}, K^{184}) are obtained by the action of aqueous chromium trioxide on the alkali metal diuranate.

Hydrated thorium(IV) and neptunium(IV) perrhenates, $M(ReO_4)_4 \cdot x\,H_2O$, are obtained from aqueous solutions of the quadrivalent actinides; the anhydrous compounds are prepared by the action of Re_2O_7 vapour on the actinide dioxides at high temperature in a sealed tube[185]. Other compounds, such as plutonium(IV) tungstate[186], are precipitated from aqueous acid media but their compositions appear to be somewhat uncertain. They are not, therefore, discussed here.

18. REFERENCES

1   S.P. Mushran, *Nature*, 158 (1946) 95.
2   F. Weigel and N. ter Meer, *Inorg. Nucl. Chem. Letters*, 3 (1967) 403.
3   C. Keller and D. Fang, *Radiochim. Acta*, 11 (1969) 123.
4   V.B. Dedov, V.V. Volkov, B.A. Gvozdev, V.A. Ermakov, I.A. Lebedev,
    V.M. Razbitnoi, P.S. Trukhlyaev, Yu. T. Chuburkov and G.N. Yakovlev,
    Radiokhimiya, 7 (1965) 453.
5   D.S. Bharadwaj and A.R. Vasudeva Murthy, *Indian J. Chem.*, 2 (1964) 391.
6   E.Chauvenet, *Compt. Rend.*, 153 (1911) 66.
7   B. Sahoo and D. Patnaik, *Nature*, 185 (1960) 683.
8   H.T. Evans Jr. and C. Frondel, *Am. Mineralogist*, 35 (1950) 251.
8a V.A. Golovnya and G.T. Bolotova, *Russ. J. Inorg. Chem.*, 6 (1961) 1256.
9   A. Rosenheim, V. Samter and J. Davisohn, *Z. Anorg. Chem.*, 35 (1903)
    424.
10 K. Ueno and M. Hoshi, *J. Inorg. Nucl. Chem.*, 32 (1970) 3817.
11 I.I. Chernyaev, V.A. Golovina and A.K. Molodkin, *Proc. 2nd Intern. Conf.
    Peaceful Uses At. Energy* (1958), Paper P/2136, Session C.10
12 A. Rosenheim and M. Kelmy, *Z. Anorg. Chem.*, 206 (1932) 31.
13 A.D. Gel'man and L.M. Zaitsev, *Zhur. Neorgan. Khim.*, 3 (1958) 1304.
    1551.
14 D.I. Ryabchikov, M.P. Volynets, V.A. Zarinskii and V.I. Ivanov, *Zhur.
    Analit. Khim.*,18 (1963) 348.
15 T.K. Keenan and F.H. Kruse, *Inorg. Chem.*, 3 (1964) 1231.
16 J.P. Nigon, R.A. Penneman, E. Staritzky, T.K. Keenan and L.B. Asprey, *J.
    Phys. Chem.*, 58 (1954) 403.
17 V.P. Zaitseva, D.P. Alekseeva and A.D. Gel'man, *Radiokhimiya*, 10 (1968)
    669.
18 F.H. Ellinger and W.H. Zachariasen, *J. Phys. Chem.*, 58 (1954) 405.
19 D.S. Gorbenko-Germanov and R.A. Zenkova, *Russ. J. Inorg. Chem.*, 11
    (1966) 282.
20 P.D. Miller, H.A. Spray and H.P. Manger, *U.S. At. Energy Comm.*,
    (Unclassified Report AECD-2740), 1949.
21 I.I. Chernyaev, V.A. Golovnya and G.V. Ellert, *Zhur. Neorgan. Khim.*,
    1 (1956) 2726.
22 C. Frondel and R. Meyrowitz, *Am. Mineralogist*, 41 (1956) 127.
23 E. Staritzky and D. Cromer, *Analyt. Chem.*, 28 (1956) 1211.
24 C.L. Christ, J.R. Clark and H.T. Evans Jr., *Science*, 121 (1955) 472.
25 R.M. Douglass, *Analyt. Chem.*, 28 (1956) 1635.
26 M. Bachelet, E. Cheylan, M. Douis and J. Goulette, *Bull. Soc. Chim.
    France*, (1952) 55,65; (1954) 173; (1956) 441.
27 W.L. Marshall and J.S. Gill, *J. Am. Chem. Soc.*, 79 (1957) 1300.
28 J.R. Ferraro, L.I. Katzin and G. Gibson, *J. Am. Chem. Soc.*, 77 (1955)
    327.
29 J.R. Ferraro, L.I. Katzin and G. Gibson, *J. Am. Chem. Soc.*, 76 (1953)
    909.
30 T. Ueki, A. Zalkin and D.H. Templeton, *Acta Cryst.*, 20 (1966) 836.
31 J.C. Taylor, M.H. Mueller and R.L. Hitterman, *Acta Cryst.*, 20 (1966) 842.

32   G. Johansson, *Acta Chem. Scand.*, 22 (1968) 389.
33   K.W. Bagnall, P.S. Robinson and M.A.A. Stewart, *J. Chem. Soc.*, (1961) 4060.
34   J.B. Laidler, *J. Chem. Soc. (A)*, (1966) 780.
35   J.L. Drummond and G.A. Welch, *J. Chem. Soc.*, (1956) 2565.
36   E. Staritzky, *Analyt. Chem.*, 28 (1956) 2021.
37   R.J. Meyer and R. Jacoby, *Z. Anorg. Chem.*, 27 (1901) 359.
38   S. Śćavničar and B. Prodić, *Acta Cryst.*, 18 (1965) 698.
39   J.L. Ryan, *J. Phys. Chem.*, 64 (1960) 1375.
40   W.E. Keder, J.L. Ryan and A.S. Wilson, *J. Inorg. Nucl. Chem.*, 20 (1961) 131.
41   K.W. Bagnall, D. Brown, P.J. Jones and P.S. Robinson, *J. Chem. Soc.*, (1964) 2531.
42   F.A. Cotton and R. Francis, *J. Chem. Soc.*, 82 (1960) 2986.
43   P.J. Alvey, K.W. Bagnall and D. Brown, to be published.
44   B.C. Smith and M.A. Wassef, *J. Chem. Soc. (A)*, (1968) 1817.
45   D.L. Plymale, *J. Inorg. Nucl. Chem.*, 31 (1969) 236.
46   A.K. Molodkin, O.M. Ivanova, Z.V. Belyakova and L.E. Kolesnikova, *Russ. J. Inorg. Chem.*, 15 (1970) 1692.
47   A.D. Garnovskii, V.I. Minkin, O.A. Osipov, V.T. Panyushkin, L.K. Isaeva and M.I. Knyazhamskii, *Russ. J. Inorg. Chem.*, 12 (1967) 1288.
48   Mazhur-ul-Haque, C.N. Caughlan, F.A. Hart and R. van Nice, *Inorg. Chem.*, 10 (1971) 115.
49   D. Brown and P.J. Jones, *J. Chem. Soc. (A)*, (1966) 733.
50   E. Späth, *Monatsh.*, 33 (1912) 853.
51   G. Gibson and J.J. Katz, *J. Am. Chem. Soc.*, 73 (1951) 5436.
52   B. Jezowska-Trzebiatowska and B. Kedzia, *Bull. Acad. Polon. Sci., Sér. Sci. Chim.*, 10 (1962) 213.
53   C.C. Addison, H.A.J. Champ, N. Hodge and A.H. Norbury, *J. Chem. Soc.*, (1964) 2354.
54   E.H.P. Cordfunke, Thermodynamics, *Proc. Symp. (1965), Intern. At. Energy Agency*, Vienna, 2 (1966) 483.
55   M.Ye. Krevinskaia, V.D. Nikol'skii, B.G. Pozharskii and Ye. Ye. Zastenker, *Radiokhimiya*, 1 (1959) 562.
56   D. Hall, A.D. Rae and T.N. Waters, *Acta Cryst.*, 19 (1965) 389.
57   J.C. Taylor and M.H. Mueller, *Acta Cryst.*, 19 (1965) 536.
58   M.H. Mueller and N. Kent Dalley, *Natl. Bur. Std (U.S.) Spec. Publ.*, No. 301, 1967, p. 253.
59   G.A. Barclay, T.M. Sabine and J.C. Taylor, *Acta Cryst.*, 19 (1965) 205.
59a   M.S. Zivadinović, *Bull. Inst. Nucl. Sci. "Boris Kidrich" (Belgrade), Physics*, 18 (1967) 1.
60   J.L. Ryan, *J. Phys. Chem.*, 65 (1961) 1099.
61   J.I. Bullock, *J. Inorg. Nucl. Chem.*, 29 (1967) 2257.
62   E.K. Hyde, *Proc. 1st Intern. Conf. Peaceful Uses At. Energy*, Geneva, 7 (1955) 281.
63   T.H. Siddall III, *J. Am. Chem. Soc.*, 81 (1959) 4176.
64   A.K. Majumdar and R.G. Bhattacharya, *Chem. Ind.*, (1970) 95.

65   A.K. Majumdar, R.G. Bhattacharya and D.C. Bera, *Chem. Ind.*, (1971) 730.
66   V.V. Savant and C.C. Patel, *J. Inorg. Nucl. Chem.*, 31 (1969) 2319.
67   J.L. Edwards and J.A. Stritar, *Science*, New York, 142 (1963) 1651.
68   R. Rascanu, *Ann. Sci. Univ. Jassy*, 16 (1930-31) 31.
69   J.T. Barr and C.A. Horton, *J. Am. Chem. Soc.*, 74 (1952) 4430.
70   A.S. Solovkin, K.A. Rybakov and A.N. Panteleeva, *Russ. J. Inorg. Chem.*, 14 (1969) 1362.
71   A.K. Majumdar and R.G. Bhattacharya, *Sci. Cult. (Calcutta)*, 35 (1969) 271.
72   C.N. Lestas and M.R. Truter, *J. Chem. Soc. (A)*, (1971) 738.
73   I.M. Kopashova, I.K. Skutov, D.S. Umreiko and R.I. Shamanovskaya, *Russ. J. Inorg. Chem.*, 12 (1967) 1748.
74   V.P. Markov and V.V. Tsapkin, *Russ. J. Inorg. Chem.*, 4 (1959) 1030.
75   I.M. Kopashova, D.S. Umreiko and R.I. Shamanovskaya, *Zh. Prikl. Spektrozk.*, 10 (1969) 675.
76   P.S. Gentile and L.S. Campisi, *J. Inorg. Nucl. Chem.*, 27 (1965) 2291.
77   J.E. Fleming and H. Lynton, *Chem. Ind.*, (1960) 1415.
78   G. Panattoni, R. Graziani, U. Croatto, B. Zarli and G. Bombieri, *Inorg. Chim. Acta*, 2 (1968) 43.
79   A.M. Golub, V.A. Kalibabchuk and K.M. Boiko, *Russ. J. Inorg. Chem.*, 13 (1968) 1091.
79a S. Fried, F. Hagemann and W.H. Zachariasen, *J. Am. Chem. Soc.*, 72 (1950) 771.
80   C.W. Bjorklund, *J. Am. Chem. Soc.*, 79 (1957) 6347.
81   C. Keller and K.H. Walter, *J. Inorg. Nucl. Chem.*, 27 (1965) 1253.
82   P. Castel, *Compt. Rend.*, 208 (1939) 519.
83   A. Colani, *Ann. Chim. Phys.*, [8], 12 (1907) 97.
84   J. d'Ans and W. Dawihl, *Z. Anorg. Chem.*, 178 (1929) 252.
85   J. Aloy, *Thèse*, Toulouse, 1901.
86   E.S. Kitchevsky and J.C. Hindman, according to B.B. Cunningham and J.C. Hindman in G.T. Seaborg and J.J. Katz (Eds.) *The Actinide Elements*, Nat. Nucl. Energy Ser. Vol. IV-14A, McGraw-Hill, New York, 1954, Chapter 12.
87   E.L. King in G.T. Seaborg, J.J. Katz and W.M. Manning (Eds.), *The Transuranium Elements*, Nat. Nucl. Energy Ser. Vol. IV-14B McGraw-Hill New York, 1949, p. 638.
88   R.G. Denotkina and V.B. Shevchenko, *Russ. J. Inorg. Chem.*, 6 (1961) 756.
89   B. Matković, B. Prodić and M. Šljukić, *Bull. Soc. Chim. France*, (1968) 1777.
90   B. Matković, B. Kojić-Prodić, M. Šljukić, M. Topić, R.D. Willett, and F. Pullen, *Inorg. Chim. Acta*, 4 (1970) 571.
91   B. Matković, B. Prodić, M. Šljukić and S.W. Peterson, *Croat. Chim. Acta*, 40 (1968) 147.
92   Troost and Ouvrard, *Ann. Chim. Phys.*, [6], 17 (1889) 227.
93   Y. Baskin, *J. Inorg. Nucl. Chem.*, 29 (1967) 383.
94   R.M. Douglass, *Acta Cryst.*, 15 (1962) 505.
95   A. Burdese and M. Lucco Borlera, *Ann. Chim. (Rome)*, 53 (1963) 344.
96   T. Dupuis and C. Duval, *Anal. Chim. Acta*, 3 (1949) 589.

97   A. Burdese and M. Lucco Borlera, *Ann. Chim. (Rome)*, 53 (1963) 333.
98   A. Chretièn and J. Kraft, *Bull. Soc. Chim. France*, 5 (1938) 372.
99   J. Schreyer and C. Baes, *J. Am. Chem. Soc.*, 76 (1954) 354.
100   M. Dominé-Bergès, *Compt. Rend.*, 236 (1953) 2242.
101   A. Weiss, K. Hartl and U. Hofmann, *Z. Naturforsch.*, 12b (1957) 669.
102   N. Pavković, M. Branica and B. Tezak, *Croat. Chim. Acta*, 40 (1968) 117.
103   R.G. Denotkina and V.B. Shevchenko, *Russ. J. Inorg. Chem.*, 12 (1967) 42.
104   R.G. Denotkina and V.B. Shevchenko, *Russ. J. Inorg. Chem.*, 12 (1967) 1237.
105   P. Chastaing, *Bull. Soc. Chim. France*, 34 (1880) 20.
106   J. Flahaut, *Nouveau Traité de Chimie Minérale*, Vol. IX, Masson, Paris, 1963, p. 1131.
107   A. Chretièn and J. Kraft, *Compt. Rend.*, 204 (1937) 1736.
108   E. Montignie, *Bull. Soc. Chim. France*, 4 (1937) 1142.
109   A. Rosenheim and G. Trewendt, *Ber.*, 55b (1922) 1957.
110   P. Castel, *Compt. Rend.*, 208 (1939) 37.
111   G.A. Barbieri, *Atti Accad. Lincei*, 19 (1911) 642.
111a G. Alberti and M.A. Massucci, *J. Inorg. Nucl. Chem.*, 32 (1970) 1719.
112   F.J. Aloy, *Bull. Soc. Chim. France*, 21 (1899) 613.
113   J. Kraft, *Compt. Rend.*, 206 (1938) 57.
114   M.-F. le Cloarec, S. Kovacevic and R. Muxart, *Rev. Chim. Minérale*, 7 (1970) 735.
115   R. Barnard, J.I. Bullock and L.F. Larkworthy, *Chem. Commun.*, (1967) 1270.
116   L.B. Asprey, *U.S. Rept. CN-3001* (1945)
117   E. Staritzky and A.L. Truitt, in G.T. Seaborg and J.J. Katz (Eds.), *The Actinide Elements*, Natl. Nucl. Energy Ser., Vol. IV-14A, McGraw-Hill, New York, 1954, p. 797.
118   G.N. Yakovlev, D.S. Gorbenko-Germanov, R.A. Zenkova, V.M. Razbitnoi and K.S. Kazanski, *J. Gen. Chem. USSR*, 28 (1958) 2653.
119   G.R. Hall and T.L. Markin, *J. Inorg. Nucl. Chem.*, 4 (1957) 137.
120   R. Barnard, J.I. Bullock and L.F. Larkworthy, *Chem. Commun.*, (1968) 960.
121   H.H. Anderson, in G.T. Seaborg, J.J. Katz and W.M. Manning (Eds.), *The Transuranium Elements*, Nat. Nucl. Energy Ser., Vol. IV-14B, McGraw-Hill, New York, 1949, pp. 806, 810, 817.
122   V.B. Dedov, V.V. Volkov, B.A. Gvozdev, V.A. Ermakov, I.A. Lebedev, V.M. Razbitnoi, P.S. Trukhlyaev, Yu.T. Chuburkov and G.N. Yakovlev, *Radiokhimiya*, 7 (1965) 453.
123   V.F. Peretrukhin, N.N. Krot and A.D. Gel'man, *Bull. Acad. Sci., USSR, Div. Chem. Sci.*, (1967) 2473.
124   P.-T. Clève, *Bull. Soc. Chim. France*, 21 (1874) 115.
125   R.J. Meyer and A. Gumperz, *Ber.*, 38 (1905) 817.
126   G.N. Wyrouboff, *Bull. Soc. Min.*, 32 (1909) 342.
127   R.J. Meyer and H. Nachod, *Annalen*, 440 (1924) 186.
128   E.Ya. Rode and Z.S. Golovleva, *Doklady Chem.*, 181 (1968) 634.

129   T.J. LaChapelle, L.B. Magnusson and J.C. Hindman, in G.T. Seaborg,
      J.J. Katz and W.M. Manning (Eds.), *The Transuranium Elements*, Natl. Nucl.
      Energy Ser., Vol. IV-14B, McGraw-Hill, New York, 1949, p. 1097.
130   H.H. Anderson, ref. 121. p. 796.
131   P. Kierkegaard, *Acta Chem. Scand.*, 10 (1956) 599.
132   O. Hauser and F. Wirth, *Z. Anorg. Chem.*, 60 (1908) 242.
133   G. Lundgren, *Arkiv Kemi*, 4 (1952) 421.
134   G. Lundgren, *Arkiv Kemi*, 2 (1951) 535.
135   A.K. Molodkin, G.A. Skotnikova and O.M. Ivanova, *Russ. J. Inorg. Chem.*,
      10 (1965) 1453; 11 (1966) 1201.
136   L.V. Lipis, B.G. Pozharskii, M.E. Pozharskaya and V.V. Fomin, *Russ. J.
      Inorg. Chem.*, 5 (1960) 1063.
137   E.G. Arutyunyan and M.A. Porai-Koshits, *J. Struct. Chem., USSR*, 4
      (1963) 96.
138   H. Grossmann, *Z. Anorg. Chem.*, 44 (1905) 229.
139   H.H. Anderson, ref. 7, p. 724.
140   K. Ueno and M. Hoshi, *J. Inorg. Nucl. Chem.*, 33 (1971) 1765.
141   O.N. Evstaf'eva, A.K. Molodkin, G.G. Dvoryantseva, O.M. Ivanova and
      M.I. Struchkova, *Russ. J. Inorg. Chem.*, 11 (1966) 697.
142   K.W. Bagnall, D. Brown and P.J. Jones, *J. Chem. Soc.*, (1965) 176.
143   K.J. Notz, *NCLO Rept. 814* (1960).
144   C.H. Secoy, *J. Am. Chem. Soc.*, 70 (1948) 3450; 72 (1950) 3343.
145   E.H.P. Cordfunke, *J. Inorg. Nucl. Chem.*, 31 (1969) 1327.
146   E. Rimbach, *Ber.*, 37 (1904) 461.
147   K. Ueno and M. Hoshi, *J. Inorg. Nucl. Chem.*, 33 (1971) 2631.
148   E. Staritzky, D.T. Cromer and D.I. Walker, *Analyt. Chem.*, 28 (1956) 1634.
149   M. Ross and H.T. Evans, *J. Inorg. Nucl. Chem.*, 15 (1960) 338.
150   V.P. Markov and V.V. Tsapkin, *Russ. J. Inorg. Chem.*, 6 (1961) 1052.
151   R.H. Goekermann and G.W. Watt, *U.S. Rept. CN-3278* (1945).
152   Y.A. Golovnya, A.K. Molodkin and V.N. Tverdokhlebov, *Russ. J. Inorg.
      Chem.*, 12 (1967) 1254.
153   G.A. Polonnikova and K.F. Kudinova, *Russ. J. Inorg. Chem.*, 6 (1961) 781.
154   F.A. Zakharova and M.M. Orlova, *Russ. J. Inorg. Chem.*, 12 (1967) 1384, 1699.
155   D.I. Ryabchikov and V.Ya. Volkova, *Compt. Rend. Acad. Sci., USSR*,
      55 (1947) 501.
156   L.F. Nilson, *Bull. Soc. Chim. France*, 23 (1875) 494.
157   J. Meyer and E. Kasper, *Z. Anorg. Allgem. Chem.*, 155 (1926) 49.
158   R. Claude, *Ann. Chim.*, 5 (1960) 165.
159   E. Montignie, *Bull. Soc. Chim. France*, 5 (1938) 197; 6 (1939) 672.
160   E. Montignie, *Bull. Soc. Chim. France*, 2 (1935) 864.
161   O. Erämetsä, *Suom. Kem.*, 15B (1942) 1.
162   C. Rammelsburg, *Ann. Phys. Chem.*, 55 (1842) 63.
163   H.H. Anderson, ref. 121, p. 825.
164   T. Moeller and N.D. Fritz, *Analyt. Chem.*, 20 (1948) 1055.
165   J.S. Sklyarenko and T.M. Chubukova, *J. Analyt. Chem. USSR*, 18 (1963)
      425.
166   B. Weaver, *Analyt. Chem.*, 40 (1968) 1894.
167   P. Artmann, *Z. Anorg. Chem.*, 79 (1912) 327; 80 (1913) 280.
168   P. Ramamurthy and C.C. Patel, *J. Inorg. Nucl. Chem.*, 25 (1963) 310.

169   K.W. Bagnall, D. Brown and A.M. Deane, *J. Chem. Soc.*, (1962) 1655.
170   J.M. Schreyer and D. Phillips, *J. Phys. Chem.*, 60 (1956) 588.
171   E. Wilke-Dörfurt and O. Schliephake, *Z. Anorg. Allgem. Chem.*, 183 (1929) 301.
172   V.V. Savant, P. Ramamurthy and C.C. Patel, *J. Less-Common Metals*, 22 (1970) 479.
173   V. Krishnan and C.C. Patel, *J. Inorg. Nucl. Chem.*, 26 (1964) 2201.
174   V. Krishnan and C.C. Patel, *Ind. J. Chem.*, 2 (1964) 416.
175   R. Paetzold and G. Bochmann, *Z. Anorg. Allgem. Chem.*, 368 (1969) 202.
176   M.D. Joesten, *Inorg. Chem.*, 6 (1967) 1598.
177   P. Ramamurthy and C.C. Patel, *Canad. J. Chem.*, 42 (1964) 856.
178   P.C.R. Choudhury, *J. Ind Chem. Soc.*, 18 (1941) 335.
179   F. Burriell Marti and C. Barcia Goyanes, *Anales Real Soc. Espan. Fis. Quim. (Madrid) Ser. B*, 50 (1954) 281.
180   Y. Marcus and D. Cohen, *Inorg. Chem.*, 5 (1966) 1740.
181   C. Keller, *Intern. Rev. Sci.*, 7 (1972) 47; Med. and Techn. Publ. Co., Aylesbury; Butterworth, London.
182   C. Palmer, *Am. Chem. J.*, 17 (1895) 374.
183   G. Lundgren and L.G. Sillen, *Arkiv Kemi*, 1 (1949) 277.
184   J. Formanek, *Annalen*, 257 (1890) 102.
185   J.-P. Silvestre, M. Pagès and W. Freundlich, *Compt. Rend.*, C272 (1971) 1808.
186   K. Watanabe, *J. Nucl. Sci. Technol.*, 5 (1968) 260.
187   G. Gordon and H. Taube, *J. Inorg. Nucl. Chem.*, 16 (1961) 272.
188   L.P. Sokhina and A.D. Gel'man, *Russ. J. Inorg. Chem.*, 5 (1960) 487.
189   L.E. Drabkina, *Zhur. Neorgan. Khim.*, 3 (1958) 1109.
190   A.D. Gel'man, A.I. Moskvih and V.P. Zaitseva, *Soviet Radiochem.*, 4 (1962) 138.
191   P. Ramamurthy and C.C. Patel, *Ind. J. Chem.*, 3 (1955) 134.
192   W.E. Bacon and J.H. Brown, *J. Phys. Chem.*, 73 (1969) 4163.
193   A.D. Gel'man and V.P. Zaitseva, *Doklady Chem.*, 157 (1964) 805.
194   A. Ekstrom, M.S. Farrell and J.J. Lawrence, *J. Inorg. Nucl. Chem.*, 30 (1968) 660.
195   V.V. Savant and C.C. Patel, *J. Less-Common Metals*, 24 (1971) 459.
196   R. Barnard, J.I. Bullock, B.J. Gellatly and L.F. Larkworthy, unpublished work (1971).
197   P. Khodadad, *Compt. Rend.*, 254 (1962) 1079.
198   P. Khodadad, *Compt. Rend.*, 255 (1962) 1617.
199   K.I. Petrov, A.K. Molodkin, O.D. Saralidze and O.M. Ivanova, *Russ. J. Inorg. Chem.*, 14 (1969) 643.
200   V.A. Golovnya and G.T. Bolotova, *Russ. J. Inorg. Chem.*, 6 (1961) 288.
201   D.S. Gorbenko-Germanov and V.C. Klimov, *Russ. J. Inorg. Chem.*, 11 (1966) 280.
202   I.I. Kapshukov, Yu.F. Volkov, E.P. Moskvichev, I.A. Lebedev and G.N. Yakovlev, *J. Struct. Chem. (USSR)*, 12 (1971) 77.
203   M.O. Mefod'eva and A.D. Gel'man, *Radiokhimiya*, 13 (1971) 597.
204   I.L. Jenkins and M.J. Waterman, *J. Inorg. Nucl. Chem.*, 26 (1964) 131.

Chapter 9

CARBOXYLATES, XANTHATES, DITHIOCARBAMATES, ALKOXIDES, PHENOXIDES AND RELATED SULPHUR COMPOUNDS

1. INTRODUCTION

Because carboxylic acids, like nitrates, are potentially bidentate ligands, there is a good chance that the large actinide(III) and (IV) cations will form complexes of high coordination number with them. The best known of these compounds are the formates, acetates and oxalates, where the remainder of the molecule is sufficiently small to allow of easy packing in a crystal lattice. Other compounds derived from aliphatic and aromatic carboxylic acids have been recorded, but little is known about them apart from the methods used for their preparation. Xanthates and N,N-dialkyldithiocarbamates have attracted rather less interest than the simpler carboxylates, largely because of the difficulties encountered in handling them, although they also will form complexes in which the actinide metal is commonly 8-coordinate. Many of the alkoxides are appreciably volatile at quite moderate temperatures, so that they could possibly be used in actinide element separation procedures, but such usage does not appear to have occurred.

2. CARBOXYLATES

Most of the actinide compounds can be prepared as the hydrates by dissolving the appropriate oxide or hydroxide in the carboxylic acid, or in the anhydrous condition by the reaction of a halide with the anhydrous acid under reflux, either alone or in a suitable inert diluent such as benzene, the hydrogen halide being evolved. The analogous reaction with hydrated salts may yield a basic carboxylate. The hydrated compounds can often be dehydrated at quite moderate temperatures. The oxalates, and most of the dibasic acid salts, together with a few of the monobasic acid compounds, particularly the long chain ones, are insoluble in water, the hydrated salts being precipitated.

(a) Formates

The only actinide(III) formates known are the dark olive-green uranium[117] and the pink americium[1] salts, $M(HCOO)_3$. The uranium compound is obtained by treating a solution of uranium tetrachloride in anhydrous formic acid with a large excess of zinc amalgam[117]. It separates from the resulting olive-brown or red solution, the colour of which depends on the concentration of uranium(III) present, as a crystalline solid which is apparently anhydrous[117]. The americium compound is obtained by evaporating the solution that results when americium(III) hydroxide is dissolved in concentrated formic acid[1].

Thorium(IV) formate, $Th(HCOO)_4 \cdot 3 H_2O$, is prepared by treating hydrated thorium dioxide with 20–80% formic acid; it is dehydrated when washed with acetone or ether[2]. The crystal symmetry of the trihydrate is monoclinic and that of the anhydrous salt is orthorhombic[3]. The dark-green anhydrous uranium salt is precipitated when uranium tetrachloride is boiled with 100% formic acid under reflux[4]; a purer product is apparently obtained by refluxing the *tetrakis* acetylacetonate (p. 214) with 100% formic acid[7]. The basic salt, $U(OH)_2(HCOO)_2 \cdot 2 H_2O$, is formed when a hydrated uranium(IV) salt is treated in the same way[4]. The anhydrous tetraformate can also be prepared by passing hydrogen chloride through a suspension of uranium metal in formic acid, after which the filtrate is refluxed until the anhydrous compound is precipitated[5], a method which is equivalent to the reaction with uranium tetrachloride, and by heating the basic formate with formic acid[6], a result which seems to be at variance with the method used to prepare the basic compound. Three crystal modifications of the anhydrous uranium compound are known, one of which is isomorphous with the thorium compound[6].

Uranyl formate monohydrate, $UO_2(HCOO)_2 \cdot H_2O$, is formed as a yellow, crystalline solid when a solution of ammonium diuranate (probably uranyl hydroxide) in dilute formic acid is evaporated. It is dehydrated at 170° and the anhydrous compound decomposes to α-uranium trioxide at 280° in a vacuum[7].

Formatocomplexes of composition $M^I Th(HCOO)_5 (M^I = K^8, Cs^2, C_5 H_5 NH^8)$, $M^{II} Th(HCOO)_6 \cdot 2 H_2 O (M^{II} = Sr, Ba^8)$, $M_2^I Th(HCOO)_6 (M = NH_4{}^8, Cs^2)$, $Cs_3 Th(HCOO)_7$ and $Cs_4 Th(HCOO)_8$ have been recorded. The penta- and hexaformato complexes are precipitated slowly when aqueous solutions of thorium(IV) formate and the alkali or alkaline earth metal formate are mixed. The dihydrated hexaformato complexes lose both molecules of water when washed with ethanol, ether or acetone[2,8]. The caesium hexa-, hepta- and octaformato complexes were prepared by heating an aqueous solution of thorium(IV) sulphate to 70–80° and treating the hot solution with the calculated quantities of caesium sulphate and barium formate. The filtrate was allowed to cool and was then left to crystallise by slow evaporation. The hepta- and octaformato complex salts are very hygroscopic, but the others are not[2]. All of these salts are insoluble in organic solvents. Their infrared spectra have been published[9].

The only oxygen-donor complex known is the dimethyl sulphoxide (DMSO) one[10], $Th(HCOO)_4 \cdot DMSO$, prepared by treating the trihydrate with the ligand.

(b) Acetates

Anhydrous thorium acetate, $Th(CH_3COO)_4$, can be prepared by heating hydrated thorium tetranitrate with acetic anhydride[11] or by heating thorium tetrachloride with a mixture of acetic acid and acetic anhydride at[12] 130°, but a simpler method is to heat thorium dioxide with 50% acetic acid at 40–60°, when the product is precipitated immediately[13]. It is a white solid, insoluble in acetic anhydride, almost insoluble in the common organic solvents and slightly soluble (1.5 g/100 ml) in water[14]. It decomposes to the dioxide above 300°. White basic acetates of composition $Th(OH)_x (CH_3COO)_{4-x}$ are precipitated when alkali hydroxide and ammonium acetate are added to

thorium(IV) salt solutions[13]. Dark-green anhydrous uranium
tetraacetate is obtained by boiling a solution of uranium tetrachloride
in glacial acetic acid[4] or, better, by reducing uranyl acetate in solution
in glacial acetic acid with zinc or cadmium under reflux[15]. Cadmium
does not form an acetatouranate(IV) complex, and although zinc does
form such a salt, the zinc acetate is eliminated from the precipitate by
washing it with hot acetic acid[15].

Crystals of uranium tetraacetate are of monoclinic symmetry
(space group $C2/c\text{-}C_{2h}^6$) and the uranium atom is 10-coordinate[16]. The
coordination geometry is a bicapped square antiprism and the compound is an acetate-bridged polymer (Fig. 21). The thorium compound
has the same structure[17].

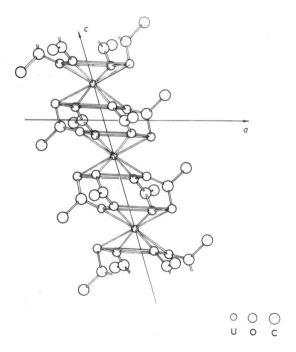

Fig. 21. Model of the polymer chain in uranium(IV) acetate (I. Jelenić, B.
Grdenić and A. Bezjak, *Acta Cryst.*, 17 (1964) 758). Reproduced by
permission of the International Union of Crystallography.

Dihydrated uranyl acetate, $UO_2(CH_3COO)_2 \cdot 2 H_2O$, crystallises from dilute acetic acid solutions of uranium trioxide. It can be dehydrated by heating to $115°$ and the anhydrous compound is obtained directly by the action of acetic anhydride on hydrated uranyl nitrate[18]. The dihydrate is a non-electrolyte in methanol[19], which suggests that the compound could be regarded as $[UO_2(CH_3COO)_2(H_2O)_2]$, a view which appears to be confirmed by a study[20] of the vibrational spectrum of the compound; from this it was deduced that the molecule is hexagonal bipyramidal with two *trans* water molecules and two *trans* bidentate acetate groups in the equatorial plane. The crystal symmetry of the dihydrate is orthorhombic, the space group being[20] $Pnc2_1$ and not $Pbn2_1$ as previously reported[21].

Hexaacetato complexes derived from thorium and uranium tetraacetates are known; the guanidinium salt, $(CN_3H_6)_2Th(CH_3COO)_6 \cdot x H_2O$, separates when the solution obtained by dissolving freshly precipitated hydrated thorium dioxide in aqueous guanidinium acetate and acetic acid at $50-60°$ is allowed to cool in air. It is also prepared by treating thorium tetraacetate with a large excess of aqueous guanidinium acetate[13]. The analogous uranates(IV), $M^{II}U(CH_3COO)_6$, where M^{II} is Mg, Fe or Zn, are precipitated when a solution of uranyl acetate in glacial acetic acid is treated with magnesium, iron or zinc metal under reflux[15]. Infrared spectral data are available for these compounds.

Actinyl(VI) triacetato complexes, $NaM^{VI}O_2(CH_3COO)_3$ (M^{VI}, U, yellow[22]; Np, green[22]; Pu, Pink[22]; Am, lemon-yellow[23]) are precipitated from aqueous acetic acid solutions of the actinide(VI) on the addition of sodium acetate. These salts are of cubic symmetry (space group $P2_13$) and the metal atom in the uranium compound[24] is bonded to six acetate oxygen atoms in a plane hexagon, with the O=U=O group normal to the plane, so making the coordination geometry a hexagonal bipyramid. There is also spectrophotometric evidence for the formation of a tetraacetato complex, $[(C_2H_5)_4N]_2UO_2(CH_3COO)_4$, in methyl cyanide solution. Acetato complexes of the type NaM^{II} $[UO_2(CH_3COO)_3]_3 \cdot 6 H_2O$, where M^{II} is Mg, Zn or Ni, are well known in analytical chemistry where, for example, zinc uranyl acetate is commonly used as a precipitant for sodium. The sodium magnesium salt is, however, rather less soluble in water than its zinc analogue and is therefore a more sensitive reagent for sodium[25]. A thallium compound, $Tl[(UO_2)_2(CH_3COO)_5(H_2O)_2]$ is also known[19].

Triphenylphosphine oxide (TPPO) and arsine oxide (TPAsO) complexes of uranyl acetate, $UO_2(CH_3COO)_2 \cdot L_2$ and $[UO_2(CH_3COO)_2 \cdot L]_2$, are obtained by treating the acetate dihydrate with the ligand in acetone ($UO_2(CH_3COO)_2 \cdot 2$ TPAsO, $[UO_2(CH_3COO)_2 \cdot TPPO]_2$), methanol ($[UO_2(CH_3COO)_2 \cdot TPAsO]_2$) or nitromethane ($UO_2(CH_3COO)_2 \cdot 2$ TPPO). The last decomposes to the light-yellow dimeric mono-TPPO complex when recrystallisation from hot acetone or dioxane is attempted, but the other three compounds are conveniently recrystallised from nitromethane and are green-yellow. Crystals of the monomeric *bis* complexes are of monoclinic symmetry (space group $P2_1/c$) and those of the dimeric monocomplexes are triclinic[27] (space group $P\bar{1}$). The uranium atom in the *bis* complexes is 8-coordinate, the environment of the uranium atom being a hexagonal bipyramid of oxygen atoms made up of two *trans* ligands and two *trans* bidentate acetate groups in the equatorial plane. In the dimer, two acetate groups bridge the two uranium atoms, and the coordination geometry about each uranium atom is close to that of a pentagonal bipyramid[26].

Orthophenanthroline (phen) and α,α dipyridyl (dipy) adducts of uranyl acetate, yellow $UO_2(CH_3COO)_2(phen) \cdot 2.5 H_2O$ and pale-green $UO_2(CH_3COO)_2(dipy)$ are precipitated from aqueous solutions of the components in the cold[29], and the light-green basic compounds, $UO_2(OH)(CH_3COO) \cdot L$ (L = phen, dipy) are precipitated from hot aqueous solutions of uranyl acetate[30]. Both pairs of compounds are insoluble in alcohols or ether.

(c) *Higher aliphatic acid compounds*

A number of thorium(IV), uranium(IV) and uranyl(VI) compounds of these acids have been recorded; for example, uranium(IV) laurate, myristate, palmitate and stearate are precipitated from aqueous solutions of uranium(IV) chloride on addition of the sodium salt of the acid. These compounds, and the corresponding propionate, are also obtained by heating solid uranium tetrachloride with a solution of the acid in benzene under reflux. They can be recrystallised from benzene[31]. Thorium citrate, $Th_3(C_6H_5O_7)_4 \cdot 7.5\ H_2O$, is precipitated when equimolar solutions of thorium nitrate and citric acid are mixed[32], and a basic thorium citrate, possibly $[Th(OH)]\ [(ThOH)_2(C_6H_5O_7)_3]$, has also been described[33].

Uranium(V) lactates U_2L_5 and $UL_2(HL)$, where H_2L is lactic acid, are obtained as brown solids by heating uranium pentaethoxide with the appropriate amount of lactic acid (2 hydroxypropanoic acid, $CH_3CH(OH)COOH$) in dry benzene[110], in which they are insoluble.

Uranyl salts of branched or straight chain carboxylic acids are quite well known; for example, the pivalate, $UO_2\ [(CH_3)_3C \cdot COO]_2$, is obtained by heating uranyl acetate with pivalic acid under reflux[28] and complex propionates of the types $KUO_2(C_2H_5COO)_3$ and $NH_4(UO_2)_2(C_2H_5COO)_5 \cdot 2\ H_2O$ have been recorded[34]. Amide adducts of uranyl carboxylates, $UO_2(RCOO)_2 \cdot R'CONR''_2$ (R = iC_3H_7, $(C_2H_5)_2CH$, $C_6H_5CH_2$; R' = H, R'' = CH_3 and R' = CH_3, R'' = iC_3H_7) are prepared by mixing aqueous, ethanolic or dichloromethane solutions of the components and adding a non-polar solvent to precipitate the complex. All of these adducts are dimeric, whereas the analogous phen and dipy complexes, $UO_2(RCOO)_2 \cdot L$, are monomers[28].

(d) *Oxalates*

Actinide(III) oxalates, $M_2(C_2O_4)_3 \cdot x\ H_2O$, are known for actinium, plutonium, americium and curium. The actinium compound is white and is best prepared by homogeneous precipitation from a solution of actinium(III) in formic acid at 90–95°, the oxalate ion being provided by the hydrolysis of dimethyl oxalate[35]. The degree of hydration is uncertain. The dark green plutonium[36], pink americium[1, 36] and white curium[37] compounds are decahydrates; they are precipitated from aqueous acid solutions of the tervalent elements at $80°$ (Pu, Am) or room temperature (Cm) on addition of oxalic acid. The plutonium compound is isomorphous[38] with $La_2(C_2O_4)_3 \cdot 10\ H_2O$. The plutonium[39] and curium[37] compounds lose water stepwise on heating, yielding the anhydrous oxalates at $300°$ and $280°$ respectively. The anhydrous compounds decompose, forming carbonates, above about $450°$ (Am[1]) or $360°$ (Cm).

Actinide(IV) oxalate hexahydrates, $M(C_2O_4)_2 \cdot 6\ H_2O$, where M is thorium[40] (white), uranium[41] (dark green), neptunium[42] (green) or plutonium[36] (green), are precipitated when oxalic acid is added to a solution of the actinide(IV) in hot, dilute hydrochloric acid (U, Pu) or in dilute nitric acid at room temperature (Th, Np). The thorium compound is also obtained when the stoicheiometric quantity of oxalic acid is added to an ethanolic solution of a hydrated thorium(IV) compound, but with an excess of oxalic acid a product of composition $H_2Th_2(C_2O_4)_5 \cdot 9\ H_2O$ is formed[40]. Homogeneous precipitation of the thorium compound, with the hydrolysis of dimethyl oxalate as the source of oxalic acid, gives a more crystalline product[43]. Thorium(IV) and uranium(IV) oxalate hexahydrates lose water when heated, yielding the dihydrate at $145°$ (Th) or $125°$ (U). The anhydrous thorium compound results at $270°$, whereas uranium oxalate dihydrate yields the monohydrate at $185°$, and this last decomposes to a mixture of uranium oxides above[44] $200°$. The water molecules are even less strongly held in $Pu(C_2O_4)_2 \cdot 4\ H_2O$, for this forms the dihydrate when it is washed with ethanol or acetone and vacuum dried[45]. The dihydrated plutonium compound decomposes when heated; the product obtained when it is heated in an inert atmosphere is probably[46] $Pu_2(C_2O_4)_3$.

Tetraoxalato complexes of the type $M_4^I M^{IV}(C_2O_4)_4 \cdot x\ H_2O$ (M^I = Na, K, NH_4) are known for thorium[40] (white), uranium[47] (near white) and plutonium[48] (two forms, green-yellow and red). They are prepared by dissolving the hydrated oxalate, $M(C_2O_4)_2 \cdot 6\ H_2O$, in a solution of alkali metal or ammonium oxalate, and crystallise when the solution is allowed to evaporate slowly in air (Th) or when ethanol is added (U, Pu). The barium uranium compound, $Ba_2U(C_2O_4)_4 \cdot 9 H_2O$, is precipitated[47] when barium chloride is added to a solution of $U(C_2O_4)_2 \cdot 6 H_2O$ in aqueous potassium oxalate; this is apparently a pale-violet solid, which is unusual for a uranium(IV) compound. Ammonium salts of what appear to be pentaoxalato complexes of thorium[40](IV) and plutonium[48](IV) are also known, as well as a potassium salt of the plutonium analogue[48]. However, a subsequent investigation of the $Pu(C_2O_4)_2$–aqueous $K_2C_2O_4$ system indicates[49] that there is no evidence of the formation of complexes higher than the tetraoxalato-species. Mixed oxalatocarbonato complexes, such as the uranium(IV) hydroxoderivative[50], described as $Na_4[U_2(OH)_4(CO_3)_3C_2O_4)(H_2O)_4]$, which is precipitated when uranium(IV) oxalate is treated with concentrated sodium carbonate solution, and the plutonium[51] compound of composition $Na_4Pu(CO_3)_2(C_2O_4)_2 \cdot 3 H_2O$, have been recorded, but the nature of these species is uncertain.

Thorium(IV) oxalate hexahydrate forms a complex with dimethyl sulphoxide (DMSO), $Th(C_2O_4)_2(DMSO)_2 \cdot H_2O$, when treated with the free ligand[10]; the other actinide(IV) oxalates should behave in a similar manner.

Neptunium(V) acid oxalate, $NpO_2C_2O_4H \cdot 2 H_2O$, is obtained as a pale green precipitate when a solution of neptunium(V) in dilute hydrochloric acid is added to oxalic acid in t-butanol[52]. The corresponding plutonium(V) compound is not known, but the addition of ammonium oxalate followed by ethanol to a solution of plutonium(V) yields[53] the rose-red ammonium salt, $NH_4PuO_2(C_2O_4) \cdot 6 H_2O$. Oxalato complexes of neptunium(V) and americium(V) should be preparable by the same procedure.

Actinide(VI) oxalates, $MO_2(C_2O_4) \cdot 3 H_2O$, are known for uranium[36, 54] (yellow), neptunium[55] (grey-green) and plutonium[36, 56] (pink). They are precipitated when oxalic acid is added to a solution of the actinide(VI) in dilute nitric acid at room temperature (Np^{55}) or when heated (U, Pu^{36}), although heating is probably unnecessary. The solubility decreases with increasing atomic number[55]. The uranium compound loses two molecules of water at $100°$ and the third at $190°$; the anhydrous compound decomposes[57] above about $350°$, whereas $PuO_2(C_2O_4) \cdot 3\ H_2O$ decomposes explosively[56] at $180°$.

Oxalato complexes of the types $M_2^I[UO_2(C_2O_4)_2(H_2O)] \cdot x\ H_2O$ and $M_2^I[(UO_2)_2(C_2O_4)_3]$ have been reported; the former crystallise from aqueous solution at room temperature[58] and dehydrate[59] readily when heated to moderate temperatures, yielding $M_2^I[UO_2(C_2O_4)_2(H_2O)]$, where M^I is NH_4, Li–Cs or a dialkylammonium group. Crystals of $(NH_4)_2[(UO_2)_2(C_2O_4)_3]$, obtained anhydrous by evaporation of an aqueous solution of the components at $50°$, are of monoclinic symmetry (space group $P2_1/n$); all of the uranium atoms are 7-coordinate and lie at the centres of almost symmetrical pentagonal bipyramids (Fig. 22), the oxalate anions being ter- and quadridentate[60].

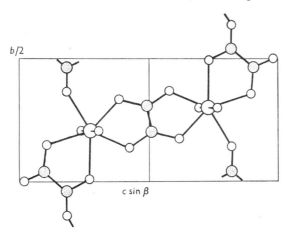

Fig. 22. Section of the chains in $(NH_4)_2(UO_2)_2(C_2O_4)_3$; the uranium atoms are shown as large circles, the carbon atoms as intermediate circles, stippled, and the oxygen atoms as small circles (N.W. Alcock, *Chem. Commun.*, (1968) 1327). Reprinted by permission of the Chemical Society, London.

Anionic oxalatourea complexes of composition
$M_2^I[UO_2(C_2O_4)_2(CO(NH_2)_2)]\cdot x\,H_2O$ are formed by displacement of
water from the aquoanions. The compounds crystallise when hot
solutions of the aquo complex, to which the calculated quantity of
urea has been added, are cooled[61]. Hydroxo complexes of the type
$M_3^I[UO_2(OH)(C_2O_4)_2]\cdot x\,H_2O$ are also known, formed by treating
the aquo complex with alkali hydroxide[62].

The only oxygen- or nitrogen-donor complexes recorded are
the N,N-dimethylformamide[63](DMF), orthophenanthroline (phen),
α,α-dipyridyl[29] (dipy) and hydrazine[64] compounds, $UO_2(C_2O_4)\cdot 3$ DMF,
$UO_2(C_2O_4)\cdot L\cdot H_2O$ (L = phen, dipy), $UO_2(C_2O_4)\cdot N_2H_4\cdot 0.75\,H_2O$
and $UO_2(C_2O_4)\cdot 2\,N_2H_4\cdot H_2O$. The DMF complex is precipitated when
oxalic acid is added to an aqueous solution of uranyl nitrate and DMF
at 60–70°, whereas the phen and dipy complexes are precipitated from
a hot, aqueous solution of uranyl oxalate on the addition of the
appropriate ligand. All of these are yellow, crystalline solids. The bright
yellow monohydrazine complex is prepared by treating a suspension of
anhydrous uranyl oxalate in ethanol with 92% hydrazine at room
temperature, a reaction which yields the yellow-brown *bis* hydrazine
complex at 50–60°.

(e) Long-chain dicarboxylates

The addition of succinic, glutaric, adipic or azelaic acid to
aqueous thorium(IV) nitrate or uranium(IV) chloride at pH 2.2–2.3
yields[65] basic thorium (white) or uranium (dark-green) compounds of
composition $(MOH)_2[(CH_2)_n(COO)_2]_3\cdot x\,H_2O$. Very little is known
about these compounds.

(f) Aromatic carboxylates

The thorium(IV) nitrobenzoates are precipitated when the acid
(o-, m-, p-) is added to an aqueous solution of thorium(IV). In contrast
to m-nitrobenzoic acid, which yields the anhydrous tetrabenzoate, the
o- and p-acids yield partially hydrolysed species[66]. A basic thorium

salicylate (sal), ThO(sal)$_2$, is precipitated from thorium(IV) solutions by salicylic acid[67] and a pyridine (py) adduct of uranium(IV) dichloride disalicylate, [UCl$_2$(sal)$_2$(py)$_2$], a greenish-white solid, is precipitated when pyridine is added to a solution of uranium tetrachloride and salicylic acid in butyl acetate[68]. Salts of a cationic uranyl complex, [UO$_2$(sal)(py)(H$_2$O)]$^+$, are also known[68].

(g) Ethylenediaminetetraacetic (H$_4$EDTA) and nitrilotriacetic (H$_3$NTA) acid complexes

The thorium(IV) and uranium(IV) compounds, M(EDTA)· 2 H$_2$O, are prepared respectively by adding thorium tetranitrate or uranium disulphate to a boiling aqueous solution of H$_4$EDTA. White crystals of the thorium compound, and green ones of the uranium analogue, separate when the resulting solution is concentrated. The anhydrous compounds are obtained by heating the dihydrate at 214° (Th) or 135° (U)[69].

The·grey neptunium(V) compound, (NpO$_2$)$_2$H$_2$EDTA·5 H$_2$O, precipitates[70] when the acid is added to a neutral solution of neptunium(V) and the pale-brown cobalt(III) hexammine salt, [Co(NH$_3$)$_6$] [NpO$_2$(EDTA)] ·3 H$_2$O is precipitated from neptunium(V) solutions at pH 7–8. One molecule of water is lost when the compound is heated to 110°.

The lemon-yellow uranyl(VI) compound, UO$_2$(H$_2$EDTA)·H$_2$O, precipitates when a mixture of H$_4$EDTA and aqueous uranyl acetate is heated to boiling[71]. The analogous greenish yellow H$_3$NTA compound, UO$_2$HNTA·5 H$_2$O, separates[72] when the solid acid is added to a boiling solution of uranyl acetate. The two compounds dehydrate at 120° and 130° respectively. All of the above are more salt-like in character than their *d*-transition element analogues; it is uncertain whether the nitrogen atoms in either of these acids are coordinated to the actinide metal atom in the compounds.

3. XANTHATES, $(ROCS_2)^-$

The red-orange uranyl complexes, $UO_2(ROCS_2)_2$ (R = CH_3, C_2H_5, n or iC_3H_7, n or iC_4H_9, *etc.*) have been isolated in poor yield by treating concentrated aqueous uranyl nitrate with the appropriate potassium alkyl xanthate. They are easily hydrolysed and are rather unstable with respect to decomposition[73]. No other actinide compounds of this type have been reported.

4. DIALKYLDITHIOCARBAMATES, $(R_2NCS_2)^-$

Nearly all of the reported actinide compounds are diethyldithio-carbamates (dtc); the main types of compound known are $M(dtc)_3(Pu)$, $(C_2H_5)_4NM(dtc)_4$ (Np, Pu), $M(dtc)_4$ (Th, U, Np, Pu), $M(dtc)_4X$ (X = Cl, Br; Pa), UO_2 (dtc)$_2$ and $M^IUO_2(dtc)_3$. In addition to these diethyl derivatives, the red-orange uranyl(VI) compounds, $UO_2(R_2NCS_2)_2$, where R is C_3H_7 or C_4H_9, and the analogous *N*-ethyl and *N*-isobutyl dithiocarbamates have been reported[74]. The uranyl(VI) compounds are appreciably more stable with respect to decomposition than are the corresponding xanthates.

Compounds of the tervalent actinides have not attracted much attention, the reported work being limited to that with neptunium and plutonium. A dark-brown solution which probably contains $Np(dtc)_3$ is obtained by treating neptunium tribromide in ethanol with the stoicheiometric quantity of Na(dtc) in the absence of oxygen and water, but the compound could not be isolated because oxidation to neptunium(IV), with the formation of $Np(dtc)_4$, occurs rapidly. In the presence of tetraethylammonium bromide, however, the dark-green complex $(C_2H_5)_4N\ Np(dtc)_4$ is formed. The analogous reactions with plutonium tribromide yield bright-green $Pu(dtc)_3$ and, in the presence of the alkylammonium halide, $(C_2H_5)_4NPu(dtc)_4$ is obtained[75]. Although $Pu(dtc)_3$ is not isostructural with any of the lanthanide *tris* chelates, the *tetrakis* anionic complexes have the same structures as their lanthanide analogues. The crystals are of monoclinic symmetry (space group $P2_1/c$) and the metal atom is 8-coordinate (Fig. 23), the polyhedron being a highly distorted dodecahedron of approximately Cs symmetry[75].

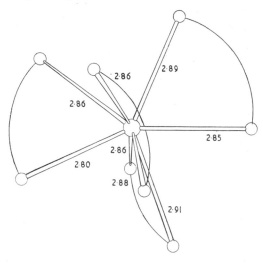

Fig. 23. The coordination polyhedron in the Np(dtc)$_4^-$ ion (D. Brown, D.G. Holah and C.E.F. Rickard, *J. Chem. Soc. (A)*, (1970) 786). Reprinted by permission of the Chemical Society, London.

Actinide(IV) *tetrakis* complexes, M(dtc)$_4$, are known[76, 77] for thorium (white), uranium (golden-yellow), neptunium (orange-red) and plutonium (black). They are prepared by treating the hexachlorometallate (IV), $M_2^I M^{IV} Cl_6$, the tetrachloride or an oxygen-donor complex, such as PuCl$_4 \cdot 3$ DMSO, with the stoicheiometric quantity of sodium[76] or lithium[77] diethyldithiocarbamate in ethanol. The residue which remains when the reaction mixture is evaporated in a vacuum is then extracted with benzene, from which the compounds are precipitated with diethyl ether[76] or petroleum ether[77]. The thorium(IV) and uranium(IV) compounds are also formed by insertion of carbon disulphide into the metal-nitrogen bond of the diethylamides[78]. The actinide(III) and actinide(IV) compounds are very readily oxidised in air, especially in ethanol solution, and are decomposed by water, so that their preparation must be carried out in a dry, inert atmosphere.

The four actinide(IV) compounds have the same structures; the crystals are of monoclinic symmetry (space group $C2/c$-C_{2h}^6) and the molecular geometry is close to an ideal dodecahedron[79]. The proton n.m.r. spectra of the thorium(IV) and uranium(IV) complexes in $CDCl_3$ or C_6D_6 show that all of the ethyl groups in these complexes are equivalent[77]. The infrared and u.v./visible spectra of the four complexes have also been published[76, 77].

Actinide(V) *pentakis* dithiocarbamates are unknown, and only the protactinium *tetrakis* diethyldithiocarbamate halides, Pa(dtc)$_4$X (X = Cl, Br, dark red[80]) are known for this oxidation state. They are prepared by treating a suspension of the pentahalide in dichloromethane with an excess of Na(dtc); they are decomposed by water and are soluble in benzene and in dichloromethane.

The red uranyl(VI) *bis* complex, $UO_2(dtc)_2$, is prepared by treating aqueous uranyl nitrate with Na(dtc) and can be recrystallised from water[81]. A purple-brown plutonyl(VI) complex appears to be formed in a similar manner, but its composition is not known[82]. The red uranyl(VI) anionic *tris* complexes, $M^I UO_2(dtc)_3(M^I = K, (CH_3)_4N)$, are prepared in the same way as the *bis* complexes, either from water[81] (K, $(CH_3)_4N$) or from anhydrous methanol[83]. These salts are soluble in a variety of organic solvents but are not very stable with respect to decomposition in solution[81]. Crystals of the tetramethylammonium salt are of orthorhombic symmetry (space group $Pna2_1$) and the anion is a distorted hexagonal bipyramid, the six sulphur atoms being staggered with respect to the hexagonal plane, to which the O=U=O group is normal[83].

The monohydrated potassium salt, $KUO_2(dtc)_3 \cdot H_2O$, reacts with triphenylphosphine oxide and with the analogous arsine oxide to form 1:1 complexes $UO_2(dtc)_2 \cdot L$. Crystals of these are of monoclinic symmetry (space group $P2_1/c$) and the uranium atom is 7-coordinate, lying at the centre of a pentagonal bipyramid[84]. The four sulphur atoms of the (dtc)$^-$ ions and the oxygen atom of the ligand lie in the pentagonal plane.

5. DIALKYLDISELENOCARBAMATES, $(R_2NCSe_2)^-$

The only recorded actinide compound of this class is the orange uranium(IV) *tetrakis* diethyldiselenocarbamate, prepared by treating a solution of uranium tetrachloride in methanol with the diethylammonium salt of the ligand. The compound melts at 112° and, like its sulphur analogue, it decomposes on exposure to air[116]. The complex is also formed by the insertion of carbon diselenide into the metal-nitrogen bonds of uranium(IV) diethylamide[78].

6. ALKOXIDES

Compounds of the type $M(OR)_4$, where M is thorium (white), uranium (light-green), neptunium (brown) and plutonium (green), are all solids. Pentaalkoxides, $M(OR)_5$, are known for uranium and these are all dark brown liquids except $U(OMe)_5$ and $U(OiPr)_5$, which are red and

TABLE 26

ACTINIDE ALKOXIDES[a] AND TRIALKYLSILYLOXIDES

$M(OR)_4$	M = Th R = Me, Et, iPr[85], nBu[86], iBu, secBu[87], nPent, neoPent[86].
	M = U R = Me, Et, tBu[88].
	M = Np R = Me, Et[89]
	M = Pu R = iPr, tBu, 3EtPent[90]
$M(OR)_5$	M = U R = Me[91,92], Et[92-94], CF$_3$CH$_2$[91], nPr[91,92], iPr[91,94], nBu[91,92], secBu, iBu, tBu[91,95], nPent[92], Pent (7 isomers), tHex (3 isomers), tHept (2 isomers)[95] SiMe$_3$, SiEtMe$_2$, SiEt$_2$Me, SiEt$_3$[96]
$M(OR)_6$	M = U R = Me, Et, nPr[97], iPr[97,98], nBu[97], secBu, tBu[98]. SiMe$_3$, SiEtMe$_2$, SiEt$_2$Me, SiEt$_3$[96]
$MO_2(OR)_2$	M = U R = Me[98], Et[98,99], nPr, iBu[98], tBu[99], iPent[98,100]

[a]Me = CH_3; Et = C_2H_5; Pr = C_3H_7; Bu = C_4H_9; Pent = C_5H_{11}; Hex = C_6H_{13}; Hept = C_7H_{15}.

brown solids respectively. However, only alkoxide halides $M(OR)_4X$ and $M(OR)_3X_2$ have been reported for neptunium(V) and the only protactinium(V) alkoxides recorded are complexes derived from the, so far, unknown diethoxytrihalide, $Pa(OEt)_2X_3$. Uranium hexa-alkoxides, $U(OR)_6$ (red liquids or solids), and uranyl(VI) dialkoxides, $UO_2(OR)_2$ (brown or red solids) are also quite well known. Uranium penta- and hexa-trialkylsilyloxides, which are brown to orange and orange to red solids respectively, have also been prepared. The principal compounds of these types are listed in Table 26.

(a) Tetraalkoxides

These are prepared by reaction of the tetrachloride with the alkali metal alkoxide dissolved in the parent alcohol[85,88-90] or with the alcohol and potassium amide in anhydrous liquid ammonia[88], and by treating the hexachlorometallate, suspended in benzene, with the appropriate alcohol and an excess of ammonia:

$$MCl_4 + 4\,M^IOR \xrightarrow{ROH} M(OR)_4 + 4\,M^ICl$$

$$UCl_4 + 4\,KNH_2 \xrightarrow{NH_3(1)} [U(NH_2)_4] + 4\,KCl$$
$$\xrightarrow{4\,ROH} U(OR)_4 + 4\,NH_3$$

$$(C_5H_5NH)_2PuCl_6 + 4\,ROH + 4\,NH_3 \rightarrow Pu(OR)_4 + 4\,NH_4Cl + 2\,C_5H_5NHCl$$

$U(OMe)_4$ and $U(OEt)_4$ have also been prepared by reaction of the diethylamide, $U[N(C_2H_5)_2]_4$, with the alcohol[88], but although the method is of general applicability, it is not very useful because of the difficulty of preparing the diethylamide.

In a number of instances the yields of the alkoxides made by these methods are poor, and in such cases it is usual to prepare these more difficultly accessible compounds by alcohol exchange from the more easily prepared tetraisopropoxide. The exchange is achieved by heating the isopropoxide with a large excess of the appropriate alcohol under reflux[85-87,90].

The uranium tetraalkoxides are very susceptible to oxidation by atmospheric oxygen, some pentaalkoxide being formed; all of the alkoxides listed in Table 26 are very easily hydrolysed and they can only be handled in a dry atmosphere. The methoxides and ethoxides (Th, U, Np) are involatile and decompose at 200° (Np) or 300° (Th, U); they are only very slightly soluble in the parent alcohol or in benzene, whereas the other tetraalkoxides sublime in a vacuum at moderate temperatures, the sublimation temperature at a given pressure increasing with the chain length of the alkyl group. Moreover, they are soluble in a variety of organic solvents. This suggests that the tetramethoxides and tetraethoxides are highly polymeric, for those thorium tetraalkoxides that are soluble in benzene have been shown to be appreciably associated. The molecular complexity n of the tetraalkoxides, $[M(OR)_4]_n$, derived from straight chain alcohols lies between 6.20 (n-butyl) and 6.44 (n-pentyl[86]), whereas the complexity of the branched chain isomers[85-87] lies between 3.8 and 4.01.

Adducts of the tetraalkoxides with the parent alcohol and with pyridine, such as the emerald green plutonium compounds[90], $Pu(OiC_3H_7)_4 \cdot iC_3H_7OH$ and $Pu(OiC_3H_7)_4 \cdot C_5H_5N$, and a few complexes of the thorium(IV) ethoxide trihalides are also known; the latter are white solids of composition $Th(OC_2H_5)Cl_3(phen)_2$ and $Th(OC_2H_5)Br_3(TPPO)_x$ ($x = 2$ or 3), where phen and TPPO are orthophenanthroline and triphenylphosphine oxide respectively. They are prepared by treating a solution of the tetrahalide in ethanol with the ligand[101].

(b) Pentaalkoxides

The uranium pentaalkoxides can be prepared by treating an alcoholic solution of the pentachloride with sodium ethoxide or ammonia, but because it is difficult to make uranium pentachloride, it is more convenient to prepare the pentaethoxide by oxidising the tetraethoxide with bromine in ethanol in the presence of sodium ethoxide and to convert this compound to other pentaalkoxides[91,92] or to the silyloxides[96] by the alcohol exchange technique. Some pentaalkoxides have been prepared by the reaction of the pyridinium salt formulated as $(C_5H_5NH)_2UOCl_5$, which is probably an equimolar mixture of $(C_5H_5NH)_2UCl_6$ and $(C_5H_5NH)_2UO_2Cl_4$, with the alcohol and ammonia in benzene. This reaction appears to yield an oxospecies, $UO(OR)_3$, which decomposes when heated in a vacuum with the formation of the pentaalkoxide[94]; it is possible that the formation of the pentaalkoxide in this way results from the oxidation of $U(OR)_4$ by $UO_2(OR)_2$.

The pentaalkoxides are soluble in the parent alcohols, carbon tetrachloride and a number of hydrocarbons such as benzene. In solution in the last, the pentamethoxide is probably present as a mixture of the dimer and the tetramer, whereas the other pentaalkoxides are dimeric in benzene[92]. The analogous silyloxides are associated to a lesser degree than the alkoxides in benzene solution.

The hexaethoxide complexes, $Na[U(OC_2H_5)_6]$, a green solid, $Ca[U(OC_2H_5)_6]_2$, a light-green solid which can be sublimed, and $Al[U(OC_2H_5)_6]_3$, a volatile green liquid, are obtained by treating the pentaethoxide with, respectively, sodium, calcium and aluminium ethoxide in benzene[102]. The sodium compound is decomposed by hydrogen chloride in ethanol with the formation of very pure $U(OC_2H_5)_5$. The pentaethoxide also forms adducts with amines, $U(OC_2H_5)_5 \cdot xL$ ($x = 2$; $(CH_3)_3N$, $(nC_3H_7)_2NH$; $x = 3$, $(CH_3)_2NH$), all of which are green liquids that are more volatile than the pentaethoxide itself[102].

Uranium pentaethoxide reacts with β-diketones and β-ketoesters with replacement of some, but not all, of the ethoxy groups[102-105]; for example, with methyl or ethylacetoacetate (L) in benzene under reflux the products are dark-brown liquids or pasty solids of composition $U(OC_2H_5)_{5-n}L_n$ ($n = 1, 2$ or 3) which can be distilled under reduced pressure[103]. In earlier work[102], however, ethylacetoacetate apparently reduced the uranium(V) compound with the formation of the uranium(IV) β-ketoester complex, although ethyltrifluoroacetate only did so when an excess of the ligand was present. The reduction may have been due to the presence of impurities.

Alkoxide halides, but not, as yet, pentaalkoxides, are known for both protactinium(V) and neptunium(V); triphenylphosphine oxide (TPPO) complexes of the diethoxide trihalide, $Pa(OC_2H_5)_2X_3 \cdot TPPO$ (X = Cl, white; Br, pale yellow) result when the ligand is added to a solution of the protactinium pentahalide in ethanol or when the pentahalide complex, $PaX_5 \cdot 2$ TPPO, is treated with hot, anhydrous ethanol. The white salts $(C_2H_5)_4N[Pa(OC_2H_5)_2X_4]$, where X is Cl or Br, result when the pentahalides are treated with the tetraalkylammonium halide in anhydrous ethanol at room temperature[106]. The uranium(V) analogues, $U(OR)_{5-n}Cl_n$ (n = 1, 2 or 3) are green liquids, prepared by treating the pentaalkoxide with hydrogen chloride[102] (R = C_2H_5, nC_3H_7) or with acetyl chloride in dry benzene[107] (R = C_2H_5, iC_3H_7). The corresponding alkoxide bromides are formed by the reaction of the pentaalkoxide with acetyl bromide[108]. Somewhat similar displacement reactions occur when uranium pentaethoxide is treated with carboxylic acids; compounds of the type $U(OC_2H_5)_{5-n}(RCOO)_n$ (R = CH_3, C_2H_5, nC_3H_7, $CH_3CH(O)$ (lactate) and C_6H_5) are formed[110], where n is 1, 2, 3 or (except for benzoic acid) 4. Most of them are brown solids or viscous liquids, but some are green solids. The uranium(V) alkoxide halides re-form the pentaalkoxide when treated with the sodium alkoxide or with the alcohol in the presence of ammonia[102]. Uranium(V) mixed halide species of composition $UCl_2Br(OR)_2$ (R = CH_3, nC_4H_9) appear to be formed by the oxidation of $UCl_2(OR)_2$ with bromine in tetrahydrofuran[109], but the information on these species is scanty.

In contrast to the behaviour of uranium tetraethoxide, the oxidation of $Np(OC_2H_5)_4$ with bromine in carbon tetrachloride, in the presence of sodium ethoxide, yields the green compound $Np(OC_2H_5)_4Br$ and not the pentaethoxide[89]; a tan coloured triethoxide dibromide, $Np(OC_2H_5)_3Br_2$ has also been reported[89].

Adducts of the pentaalkoxides with the parent alcohol and with pyridine, such as $U(OtC_4H_9)_5 \cdot tC_4H_9OH$ and $U(OR)_5 \cdot C_5H_5N$, are also known[95].

(c) Hexaalkoxides

Compounds of the type $M(OR)_6$ are known only for uranium;

the ethoxide[97] is prepared by the oxidation of $NaU(OC_2H_5)_6$ with benzoyl peroxide in ethanol; lead tetraacetate and bromine are less effective oxidants for this reaction. Although it is possible to make other uranium hexaalkoxides in the same way, they[97], and the hexasilyloxides[96], are best prepared from the ethoxide by the alcohol exchange method.

These compounds, like the pentaalkoxides, sublime at moderate temperatures in a vacuum, but they are less stable to heat than the former, and many of them decompose below $100°$. The isopropoxide[97] and the hexasilyloxides[96] are the most stable of these compounds in this respect.

The hexaalkoxides are readily soluble in the parent alcohols, benzene and other hydrocarbons; they are monomeric in benzene, apart from the isopropoxide, which seems to exhibit a slight tendency to dimerise[98], and are sufficiently powerful as oxidising agents to convert the corresponding tetraalkoxide to the pentaalkoxide[97], in contrast to the pentahalides (p. 109) which disproportionate.

(d) Uranyl(VI) alkoxides

Compounds of the type $UO_2(OR)_2$ are prepared by reaction of uranyl chloride with an alkali metal alkoxide ($Li^{98, 100}$, Na^{99}) in the parent alcohol or by alcohol exchange with the more easily preparable methoxide[99]. With some branched-chain alcohols (iC_3H_7, sec C_4H_9, tC_4H_9) disproportionation occurs when the alcohol-exchange technique is used, with the formation of $UO(OR)_4 \cdot ROH$ and $U_2O_5(OR)_2 \cdot 2\,ROH$. The former yields the hexa-alkoxide on heating[98].

The uranyl alkoxides are only sparingly soluble in alcohols and other organic solvents; alcohol adducts, such as $UO_2(OC_2H_5)_2 \cdot 2\,C_2H_5\,OH$, have been identified[98], but little else is known about these compounds. Neptunium and plutonium analogues have not been reported.

(e) Compounds derived from diols

The only compounds known of this type are derived from diketocyclobutenediol, $C_4O_2(OH)_2$, "squaric" acid, which behaves

References p. 205

very much like oxalic acid, as might be expected. The yellow,
sparingly soluble uranyl compound[111], $UO_2(C_4O_4) \cdot xH_2O$
$(1 < x < 2)$, is precipitated when squaric acid is added to a solution of
a uranyl salt at about pH 2.5, and the white thorium(IV) compound,
$Th(C_4O_4)_2 \cdot x\ H_2O$, is obtained in the same way[112]. In both cases the
precipitate is very finely divided and is difficult to filter.

7. PHENOXIDES

The more important actinide phenoxides are listed in Table 27.

TABLE 27

ACTINIDE PHENOXIDES

$U(OR)_4 \cdot xNH_3$ [a]	$R = C_6H_5,\ o\text{-},\ m\text{-},\ p\text{-}C_6H_4Cl,\ o\text{-},\ p\text{-}C_6H_4CH_3,\ \alpha,\beta C_{10}H_7$ [113]
$UCl_2Br(OR)_2$	$R = C_6H_5$ [109]
$UO_2(OR)_2 \cdot yC_5H_5N$ [b]	$R = C_6H_5,\ p\text{-}C_6H_4Cl,\ o\text{-},\ p\text{-}C_6H_4CH_3$ [114]

[a] $x = 1$–2.
[b] $y = 1$–3.

Apart from the uranium(V) compound, $UCl_2Br(OC_6H_5)_2$, all of these
are formed as adducts with nitrogen donors, such as ammonia ($U(OR)_4$)
or pyridine ($UO_2(OR)_2$) as a consequence of the method of
preparation.

The tetraphenoxide ammines are obtained as green solids by the
reaction of uranium tetrachloride, in solution in tetrahydrofuran, with
the phenol and ammonia:

$$UCl_4\ +\ 4\ ROH\ +\ (x+4)NH_3\ \rightarrow\ U(OR)_4 \cdot xNH_3\ +\ 4\ NH_4Cl$$

they are monomers in tetrahydrofuran[113]. Adducts with ammonia or
amines and the parent phenol, $U(OR)_4 \cdot 2\ ROH \cdot NRR_2'$ ($R = H, R' = H$ or

C_2H_5; $R = R' = C_2H_5$) result from the corresponding reaction of uranium tetrachloride, suspended in benzene, with the phenol and ammonia or the amine. The oxoproducts, $UO(OR)_2 \cdot 4$ $ROH \cdot NH_3$, are obtained by the analogous reaction with uranium oxodichloride[113].

The only uranium(V) phenoxide recorded is $UCl_2Br(OC_6H_5)_2$, prepared by treating $UCl_2(OC_6H_5)_2$ with bromine in ether; the starting material is apparently obtained by treating uranium tetrachloride with the calculated quantity of $NaOC_6H_5$ in ether[109], but it is not clear whether either phenoxide was isolated from the reaction mixture or not.

The pyridine adducts of the uranyl(VI) phenoxides are prepared by melting the phenol with anhydrous uranyl chloride or acetate and extracting the mixture with pyridine[114]. Di- and trimethylamine derivatives of composition (amine-H)$_2$UO$_2$(OR)$_4$ are obtained by adding the amine dropwise to uranyl acetate and the phenol in carbon tetrachloride[114], but it is not evident whether these are salts, as written, or adducts, perhaps $UO_2(OR)_2 \cdot 2$ $ROH \cdot (amine)_2$.

8. MERCAPTIDES

The light-green uranium(IV) mercaptides, $U(SR)_4$ ($R = C_2H_5$, nC_4H_9) are solids prepared by treating the *tetrakis* diethylamide with the appropriate mercaptan in ether; they are spontaneously inflammable in air and are decomposed by water or by heat[88].

The anionic maleonitriledithiolate (mnt), and isomnt, complexes of uranium(VI), $(R_4N)_2UO_2(mnt)_2$ ($R = C_2H_5, nC_3H_7$) are black solids precipitated when aqueous uranyl nitrate is treated with Na_2mnt, or Na_2imnt, and the tetraalkylammonium bromide. They are decomposed by acids or bases, and are soluble in ethanol[115]; they react with oxygen-donor ligands, L, to form[118] $(R_4N)_2$ $[UO_2(mnt)_2L]$.

9. REFERENCES

1   F. Weigel and N. ter Meer, *Z. Naturforsch.*, 25b (1971) 504.
2   V.A. Golovnya and O.M. Ivanova, *Russ. J. Inorg. Chem.*, 8 (1963) 1290.

3   M. Chevreton, B. Claudel and B. Mentzen, *J. Chim. Phys.*, 65 (1968) 890.
4   A. Rosenheim and M. Kelmy, *Z. Anorg. Chem.*, 206 (1932) 31.
5   R. Bressat, B. Claudel and A. Navarro, *Bull. Soc. Chim. France*, (1968) 4807.
6   R. Bressat, B. Claudel and A. Navarro, *Compt. Rend.*, C266 (1968) 1260.
7   G.D. Buttress and M.A. Hughes, *J. Chem. Soc. (A)*, (1968) 1272.
8   R. Weinland and A. Sark, *Ber.*, 59 (1926) 471.
9   K.I. Petrov, A.K. Molodkin, O.D. Saralidze and O.M. Ivanova, *Russ. J. Inorg. Chem.*, 13 (1968) 1716.
10   A.K. Molodkin, O.M. Ivanova, Z.V. Belyakova and L.E. Kolesnikova, *Russ. J. Inorg. Chem.*, 15 (1970) 1692.
11   S. Panda and D. Patnaik, *J. Ind. Chem. Soc.*, 33 (1956) 877.
12   R.N. Kapoor, K.C. Pande and R.C. Mehrotra, *J. Ind. Chem. Soc.*, 35 (1958) 157.
13   A.K. Molodkin, O.M. Ivanova and G.A. Skotnikova, *Russ. J. Inorg. Chem.*, 12 (1967) 57.
14   K.N. Kovalenko, D.V. Kazachenko and O.N. Samsonova, *Russ. J. Inorg. Chem.*, 8 (1963) 1163.
15   D. Grdenić and B. Korpar-Čolig, *J. Inorg. Nucl. Chem.*, 30 (1968) 1751.
16   I. Jelenić, D. Grdenić and A. Bezjak, *Acta Cryst.*, 17 (1964) 758.
17   A.A. Eliseev, A.K. Molodkin and O.M. Ivanova, *Russ. J. Inorg. Chem.*, 12 (1967) 1507.
18   E. Späth, *Monatsh.*, 33 (1912) 235.
19   G.A. Golovnya and L.K. Shubochkin, *Russ. J. Inorg. Chem.*, 8 (1963) 579.
20   B. Mentzen and G. Giorgio, *J. Inorg. Nucl. Chem.*, 32 (1970) 1509.
21   V. Amirthalingam, D.V. Chandran and V.M. Padmanabhan, *Acta Cryst.*, 12 (1959) 821.
22   J.L. Ryan and W.E. Keder, *Advan. Chem. Ser.*, 71 (1967) 335.
23   L.B. Asprey, S.E. Stephanou and R.A. Penneman, *J. Am. Chem. Soc.*, 73 (1951) 5715.
24   W.H. Zachariasen and H.A. Plettinger, *Acta Cryst.*, 12 (1959) 526.
25   R.G. Monk, *Talanta*, 15 (1968) 1259.
26   C. Panattoni, R. Graziani, G. Bandoli, B. Zarli and G. Bombieri, *Inorg. Chem.*, 8 (1969) 320.
27   G. Bandoli, R. Graziani and B. Zarli, *Acta Cryst.*, B24 (1968) 1129.
28   T.H. Siddall III, W.E. Stewart and R.L. McDonald, *Inorg. Nucl. Chem. Letters*, 5 (1969) 101.
29   V.P. Markov and V.V. Tsapkin, *Russ. J. Inorg. Chem.*, 4 (1959) 1030.
30   V.P. Markov and V.V. Tsapkin, *Russ. J. Inorg. Chem.*, 6 (1961) 1052.
31   A.M. Bhandari and R.N. Kapoor, *Z. Anorg. Allgem. Chem.*, 346 (1966) 332.
32   N.A. Skorik, V.N. Kurnok and V.V. Serebrennikov, *Soviet Radiochem.*, 9 (1967) 499.
33   M. Bobtelsky and B. Graus, *J. Am. Chem. Soc.*, 76 (1954) 1536.
34   E. Rimbach, *Ber.*, 37 (1904) 461.

35 D. Butterfield and R. Woolatt, *J. Inorg. Nucl. Chem.*, 30 (1968) 801.
36 E. Staritzky and A.L. Truitt, in G.T. Seaborg and J.J. Katz (Eds.), *The Actinide Elements*, Natl. Nucl. Energy Ser. Vol. IV–14A, McGraw-Hill, New York, 1954, p. 797.
37 V. Scherer and M. Fochler, *J. Inorg. Nucl. Chem.*, 30 (1968) 1433.
38 D.M. Chackraburtty, *Acta Cryst.*, 16 (1963) 834.
39 G.S. Rao, M.S. Subramanian and G.A. Welch, *J. Inorg. Nucl. Chem.*, 25 (1963) 1293.
40 A.K. Molodkin, T.A. Balakaeva and A.N. Kuchumova, *Russ. J. Inorg. Chem.*, 13 (1968) 1095.
41 L.E. Marchi, in L.F. Audrieth (Ed.), *Inorganic Syntheses*, Vol. 3, McGraw-Hill, New York, 1950, p. 166.
42 T.J. la Chapelle, L.B. Magnusson and J.C. Hindman, in G.T. Seaborg, J.J. Katz and W.M. Manning (Eds.), *The Transuranium Elements*, Natl. Nucl. Energy Ser., Vol. IV–14B McGraw-Hill, New York, 1949, p. 1097.
43 W.W. Wendlandt, *Analyt. Chem.*, 30 (1958) 58.
44 W.W. Wendlandt, T.D. George and G.R. Horton, *J. Inorg. Nucl. Chem.*, 17 (1961) 273.
45 I.L. Jenkins, F.H. Moore and M.J. Waterman, *Chem. Ind. (London)*, (1963) 35.
46 A. Glasner, *J. Inorg. Nucl. Chem.*, 26 (1964) 1475.
47 V. Kohlschütter and H. Rossi, *Ber.*, 34 (1901) 1472.
48 A.D. Gel'man and L.P. Sokhina, *Zhur. Neorgan. Khim.*, 3 (1958) 1100.
49 V.A. Matyukha, N.N. Krot and A.D. Gel'man, *Russ. J. Inorg. Chem.*, 13 (1968) 1099.
50 V.A. Golovnya and G.T. Bolotova, *Russ. J. Inorg. Chem.*, 6 (1961) 1259.
51 A.D. Gel'man and L.M. Zaitsev, *Zhur. Neorgan. Khim.*, 3 (1958) 1555.
52 G. Gibson, D.M. Gruen and J.J. Katz, *J. Am. Chem. Soc.*, 74 (1952) 2103.
53 A.D. Gel'man and V.P. Zaitseva, *Doklady Chem.*, 157 (1964) 805.
54 A. Colani, *Bull. Soc. Chim. France*, 37 (1925) 858.
55 M.P. Mefod'eva, N.N. Krot, T.V. Smirnova and A.D. Gel'man, *Soviet Radiochem.*, 11 (1969) 187.
56 A.D. Gel'man and L.E. Drabkina, *Zhur. Neorgan. Khim.*, 3 (1958) 1105.
57 V.M. Padmanabhan, S.C. Saraiya and A.K. Sundaram, *J. Inorg. Nucl. Chem.*, 12 (1960) 356.
58 R.N. Shchelokov, V.I. Belomestnykh and Yu. N. Mikhailov, *Russ. J. Inorg. Chem.*, 14 (1969) 829.
59 R.N. Shchelokov and V.I. Belomestnykh, *Russ. J. Inorg. Chem.*, 14 (1969) 1491.
60 N.W. Alcock, *Chem. Commun.*, (1968) 1327.
61 R.N. Shchelokov and V.I. Belomestnykh, *Russ. J. Inorg. Chem.*, 12 (1967) 843.
62 R.N. Shchelokov and V.I. Belomestnykh, *Russ. J. Inorg. Chem.*, 15 (1970) 70.
63 A.S. Solavkin and E.M. Indikov, *Russ. J. Inorg. Chem.* 13 (1968) 1051.

64   G.N. Volchkov, E.I. Krylov, V.A. Sharov and A.A. Krakhalov, *Russ. J. Inorg. Chem.*, 15 (1970) 21.
65   S.A. Merkusheva, N.A. Skorik and V.V. Serebrennikov, *Radiokhimiya*, 10 (1968) 731.
66   K.N. Kovalenko and L.V. Lutsenko, *Russ. J. Inorg. Chem.*, 12 (1967) 1582.
67   O.E. Zviagintzev and B.N. Sudarikov, *Zhur. Neorgan. Khim.*, 2 (1957) 128.
68   J.T. Barr and C.A. Horton, *J. Am. Chem. Soc.*, 74 (1952) 4430.
69   H. Brintzinger, H. Thiele and U. Müller, *Z. Anorg. Allgem. Chem.*, 251 (1943) 285.
70   A.D. Gel'man, V.I. Blokhin, T.N. Bukhtiyarova and N.N. Krot, *Russ. J. Inorg. Chem.*, 15 (1970) 975.
71   H. Brintzinger and G. Hesse, *Z. Anorg. Allgem. Chem.*, 249 (1942) 113.
72   H. Brintzinger and G. Hesse, *Z. Anorg. Allgem. Chem.*, 249 (1942) 299.
73   L. Malatesta, *Gazzetta*, 69 (1939) 408.
74   L. Malatesta, *Gazzetta*, 69 (1939) 752.
75   D. Brown, D.G. Holah and C.E.F. Rickard, *J. Chem. Soc. (A)*, (1970) 786.
76   K.W. Bagnall, D. Brown and D.G. Holah, *J. Chem. Soc. (A)*, (1968) 1149.
77   J.P. Bibler and D.G. Karracker, *Inorg. Chem.*, 7 (1968) 982.
78   K.W. Bagnall, and E. Yanir, unpublished observations.
79   D. Brown, D.G. Holah, C.E.F. Rickard and P.T. Moseley, *U.K. At. Energy Authority Rept.*, R-6907 (1971).
80   P.R. Heckley, D.G. Holah and D. Brown, *Canad. J. Chem.*, 49 (1971) 1151.
81   R.A. Zingaro, *J. Am. Chem. Soc.*, 78 (1956) 3568.
82   B.G. Harvey, H.G. Heal, A.G. Maddock and E.L. Rowley, *J. Chem. Soc.*, (1947) 1010.
83   K. Bowman and Z. Dori, *Chem. Commun.*, (1968) 636.
84   R. Graziani, B. Zarli, A. Cassol, G. Bombieri, E. Forsellini and E. Tondello, *Inorg. Chem.*, 9 (1970) 2116.
85   D.C. Bradley, M.A. Saad and W. Wardlaw, *J. Chem. Soc.*, (1954) 1091.
86   D.C. Bradley, A.K. Chatterjee and W. Wardlaw, *J. Chem. Soc.*, (1956) 2260.
87   D.C. Bradley, A.K. Chatterjee and W. Wardlaw, *J. Chem. Soc.*, (1956) 3469.
88   R.G. Jones, G. Karmas, G.A. Martin Jr. and H. Gilman, *J. Am. Chem. Soc.*, 78 (1956) 4285.
89   E.T. Samulski and D.G. Karracker, *J. Inorg. Nucl. Chem.*, 29 (1967) 993.
90   D.C. Bradley, B. Harder and F. Hudswell, *J. Chem. Soc.*, (1957) 3318.
91   R.G. Jones, E. Bindschadler, G. Karmas, G.A. Martin, Jr., J.R. Thirtle, F.A. Yoeman and H. Gilman, *J. Am. Chem. Soc.*, 78 (1956) 4289.
92   D.C. Bradley and A.K. Chatterjee, *J. Inorg. Nucl. Chem.*, 4 (1957) 279.
93   R.G. Jones, E. Bindschadler, G. Karmas, F.A. Yoeman and H. Gilman, *J. Am. Chem. Soc.*, 78 (1956) 4287.
94   D.C. Bradley, B.N. Chakravarti and A.K. Chatterjee, *J. Inorg. Nucl. Chem.*, 3 (1957) 367.
95   D.C. Bradley, R.N. Kapoor and B.C. Smith, *J. Chem. Soc.*, (1963) 1023.
96   D.C. Bradley, R.N. Kapoor and B.C. Smith, *J. Chem. Soc.*, (1963) 204.

97   R.G. Jones, E. Bindschadler, D. Blume, G. Karmas, G.A. Martin Jr.,
     J.R. Thirtle, F.A. Yoeman and H. Gilman, *J. Am. Chem. Soc.*, 78 (1956)
     6030.
98   D.C. Bradley, A.K. Chatterjee and A.K. Chatterjee, *J. Inorg. Nucl. Chem.*,
     12 (1959) 71.
99   R.G. Jones, E. Bindschadler, G.A. Martin Jr., J.R. Thirtle and H. Gilman,
     J. Am. Chem. Soc., 79 (1957) 4921.
100 H. Albers, M. Deutsch, W. Krastinat and H. von Osten, *Ber.*, 85 (1952) 267.
101 B.C. Smith and M.A. Wassef, *J. Chem. Soc. (A)*, (1968) 1817.
102 R.G. Jones, E. Bindschadler, D. Blume, G. Karmas, G.A. Martin Jr.,
     J.R. Thirtle and H. Gilman, *J. Am. Chem. Soc.*, 78 (1956) 6027.
103 A.M. Bhandari and R.N. Kapoor, *Canad. J. Chem.*, 44 (1966) 1468.
104 A.M. Bhandari and R.N. Kapoor, *Austral. J. Chem.*, 20 (1967) 233.
105 S. Dubey, S.N. Misra and R.N. Kapoor, *Z. Naturforsch.*, 25b (1970) 476.
106 D. Brown and C.E.F. Rickard, *J. Chem. Soc. (A)*, (1971) 81.
107 A.M. Bhandari and R.N. Kapoor, *J. Chem. Soc. (A)*, (1967) 1618.
108 A.M. Bhandari and R.N. Kapoor, *J. Prakt. Chem.*, 35 (1967) 284.
109 *Ethyl Corporation Report TID-19367* (1963).
110 S. Dubey, A.M. Bhandari and R.N. Kapoor, *Z. Naturforsch.*, 25b (1970) 18.
111 P.H. Tedesco and H.F. Walton, *Inorg. Chem.*, 8 (1969) 932.
112 K.W. Bagnall, unpublished observations.
113 H. Funk and K. Andrä, *Z. Anorg. Allgem. Chem.*, 361 (1968) 199.
114 H. Funk and K. Andrä, *Z. Anorg. Allgem. Chem.*, 362 (1968) 93.
115 L. Zimmer and K.H. Lieser, *Inorg. Nucl. Chem. Letters*, 7 (1971) 563.
116 B. Lorenz, R. Kirmse and E. Hoyer, *Z. Anorg. Allgem. Chem.*, 378 (1970)
     144.
117 B. Jezowska-Trzebiatowska and J. Drozdzynski, *J. Inorg. Nucl. Chem.*,
     31 (1969) 727.
118 L. Zimmer and K.H. Lieser, *Inorg. Nucl. Chem. Letters*, 7 (1971) 1163.

Chapter 10

COMPLEXES WITH β-DIKETONES, TROPOLONE, 8-HYDROXYQUINOLINE AND OTHER CHELATING LIGANDS

1. INTRODUCTION

The β-diketones, the analogous β-ketoesters and tropolone (2-hydroxy-2,4,6-cycloheptatrien-1-one) are short "bite" oxygen-donor chelating ligands which give rise to actinide complexes in which the metals have coordination numbers of 8 or even higher. The β-diketone complexes are more soluble in organic than in an aqueous media, and the ligands can therefore be used to extract the actinide from aqueous solution under the right conditions. This applies also to 8-hydroxy-quinoline and to cupferron (*N*-nitroso-*N*-phenylhydroxylamine), both of which have also found some application to the analytical chemistry of the actinides. Another useful property of the actinide(III) and actinide(IV) β-diketone and β-ketoester complexes is their volatility at quite moderate temperatures when under low pressure; this suggests that it might be possible to separate these elements by the fractional sublimation of the complexes formed with these ligands. The few known Schiff's-base and phthalocyanine complexes are also discussed in this chapter.

2. β-DIKETONE AND β-KETOESTER COMPLEXES

A large number of the uranium(IV) complexes, UL_4, of these ligands, $RCOCH_2COR'$, in which the groups R and R' have been varied over a wide range, were prepared in the course of an investigation into the change of the volatility of the complexes with the nature of the substituents R and R' in the ligand[1-3]. The ligands used are given in Table 28. Several thorium(IV)[4] and uranyl(VI)[5] complexes of other,

TABLE 28

URANIUM(IV) β-DIKETONE, β-KETOESTER AND β-KETOALDEHYDE COMPLEXES, UL_4^{1-3}

Ligands $RCOCH_2COR'$		
β-diketones	$R = CH_3$	$R' = CH_3, C_2H_5, n, iC_3H_7, n, tC_4H_9,$ $nC_5H_{11}, nC_6H_{13}, C_6H_5, CH_2OC_2H_5,$ 2-furyl
	$R = CF_3$	$R' = CH_3, CF_3, C_2H_5, n, iC_3H_7, n, i,$ $tC_4H_9, nC_5H_{11}, C_6H_5$
	$R = C_2H_5$	$R' = C_2H_5$
	$R = nC_3H_7$	$R' = nC_3H_7$
	$R = C_6H_5$	$R' = C_6H_5, o, m, pCH_3OC_6H_4, o, m,$ $pC_6H_4NO_2, m, pC_6H_4NH_2$
β-ketoesters	$R = CH_3,$	$R' = OC_2H_5, OnC_4H_9$
	$R = CF_3,$	$R' = OCH_3, OC_2H_5, OnC_4H_9$
β-ketoaldehydes	$R = CH_3$ or C_6H_5	$R' = H$

more complicated, kinds of β-diketone are also known. In most instances the ligands have been used to prepare the complex of a single element only, usually of thorium(IV), uranium(IV) or uranyl(VI), but some of the ligands have been applied to a number of the actinide elements, or to a particular element in more than one oxidation state; the discussion that follows is restricted to such instances. The compounds under consideration are given in Table 29.

These complexes are easy to prepare, the usual method being to add the ligand to an aqueous solution of the actinide in the appropriate oxidation state, adjusting the pH of the solution with alkali until the complex is precipitated.

(a) Actinide(III) complexes

The pale-rose americium(III) *tris* complexes are made by the general procedure, given above, using the ligand ((H)AA[6], (H)THD[7]) or the ammonium salt ((H)BTA, (H)TTA[6]); when an excess of the

TABLE 29

ACTINIDE(III), ACTINIDE(IV) AND URANYL(VI) β-DIKETONE COMPLEXES[a]

Am(AA)$_3$·H$_2$O	M(AA)$_4$ (M = Th, U, Np, Pu)	Pa(AA)$_2$Cl$_3$	UO$_2$(AA)$_2$
CsM(HFAA)$_4$ (M = Am, Cm)	U(HFAA)$_4$		
	M(BA)$_4$ (M = Th, U, Pu)		UO$_2$(BA)$_2$
	M(PHD)$_4$ (M = Th, U)		
Am(BTA)$_3$·3 H$_2$O	U(BTA)$_4$		
Am(TTA)$_3$·3 H$_2$O	M(TTA)$_4$ (M = Th, U, Np[b], Pu)		UO$_2$(TTA)$_2$·2 H$_2$O
Am(THD)$_3$	M(THD)$_4$ (M = Th, U)		
	M(FOD)$_4$ (M = Th, U, Np)		

[a] Abbreviations: (H)AA, acetylacetone (pentane-2,4-dione); H(HFAA), hexafluoro-acetylacetone (1,1,1,5,5,5-hexafluoropentane-2,4-dione); H(BA), benzoylacetone (1-phenylbutane-1,3-dione); H(PHD), benzoyltrifluoroacetone (1-phenyl-4,4,4-trifluorobutane-1,3-dione); H(DBM), dibenzoylmethane (1,3-diphenylpropane-1,3-dione); (H)TTA, thenoyltrifluoroacetone (1,1,1-trifluoro-4-α-thienylbutane-2,4-dione); H(THD), dipivaloylmethane (2,2,6,6-tetramethylheptane-3,5-dione); H(FOD), 1,1,1,2,2,3,3-heptafluoro-7,7-dimethyloctane-4,6-dione.
[b] Identification uncertain.

caesium salt, CsHFAA, is used, the product is a yellow hydrated anionic complex, $CsAm(HFAA)_4 \cdot H_2O$, from which the pale-rose anhydrous compound is obtained on recrystallisation from 1-butanol[8]. The hydrated curium compound, $CsCm(HFAA)_4 \cdot H_2O$, is prepared in the same way as the americium one[9], but it is not clear whether this is dehydrated on recrystallisation or not.

The crystal structures of some of these compounds are known; $Am(THD)_3$ is of monoclinic symmetry (space group $P2_1/c$) and is isomorphous[7] with $Pr(THD)_3$ and $Nd(THD)_3$. The structure of the lanthanide compounds is made up of dimers, $[M(THD)_3]_2$, in which each metal atom is surrounded by 7 oxygen atoms[7a]. In contrast to this, the americium atom in $CsAm(HFAA)_4$, crystals of which are of orthorhombic symmetry (space group $Pbcn$), is 8-coordinate; the geometry is a dodecahedron of approximately D_2 symmetry[8] (Fig. 24).

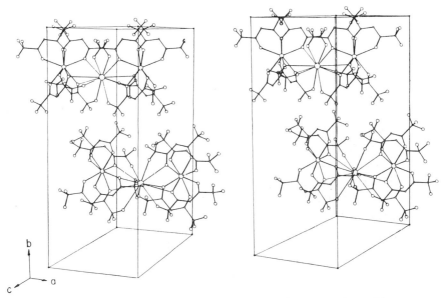

Fig. 24. A portion of the structure of $CsEu(HFAA)_4$, which is isomorphous with $CsAm(HFAA)_4$, drawn as a stereoscopic pair. Two anion-cation chains are shown and the circles represent, in decreasing order of size, Cs, Eu, O, F and C respectively (J.H. Burns and M.D. Danford, *Inorg. Chem.*, 8 (1969) 1780). Copyright (1969) by the American Chemical Society and reprinted by permission of the copyright holder.

References p. 227

Crystals of the monohydrated complex[7] are isomorphous with those of the analogous curium compound[9], but the coordination geometry has not been reported.

The americium(III) complexes, $Am(AA)_3$, $Am(BTA)_3$ and $Am(TTA)_3$, decompose to americium dioxide at 200-400°, but $Am(THD)_3$ sublimes[7] at 124-135°/10^{-5} torr and melts at 216-218° in a sealed tube. $CsAm(HFAA)_4$ also sublimes at 130-140°/10^{-6} torr[7]; the reported sublimation of the monohydrate may be the result of dehydration. Two of these complexes, $Am(BTA)_3$ and $Am(TTA)_3$, are reported to be soluble in chloroform and in ethanol[6], and it is probable that the other complexes will be found to be soluble in these solvents. Although similar complexes of the higher actinides have not been isolated, there is abundant evidence from solvent extraction experiments to indicate that analogous species are formed.

(b) Actinide(IV) complexes

Apart from the complexes derived from (H)TTA (Th, U^{10}, Np^{45}, Pu^{10}), which precipitate when the ligand is added to an aqueous solution of the actinide(IV) without any adjustment of the pH, the complexes are made by the general method given earlier (p. 211). The neptunium complex, $Np(AA)_4$, has also been isolated by extracting it from aqueous solution (pH 4.5-5.0) into benzene and evaporating the resulting solution[16]. The thorium compounds ((H)AA[12], (H)BA[4], (H)PHD[13], (H)TTA[10], (H)DBM[4], (H)THD, (H)FOD[11]) range from white to creamy-yellow in colour, the uranium ones ((H)AA, (H)HFAA, (H)BA[3], (H)PHD[13], (H)BTA[3], (H)TTA[10], (H)DBM[3], (H)THD, (H)FOD[11]) are green to dark-brown, the neptunium ones ((H)TTA[45], (H)FOD[11]) are green and the plutonium complexes ((H)AA[14, 15], (H)BA[15], (H)TTA[10], (H)FOD[11]) are brown.

The thorium[17] and uranium[18, 19] complexes $M(AA)_4$ are dimorphic; crystals of both forms are of monoclinic symmetry, but of differing space group (α, $P2_1/c$; β, $C2/c$). The β-form of $Th(AA)_4$ transforms to the α-form at room temperature[17]. The neptunium[16] and plutonium[20] compounds have the $\beta Th(AA)_4$ structure and do not seem

to exist as the α-form. The metal atom in these complexes is 8-coordi-
nate, in the shape of a slightly distorted square antiprism in both
modifications, but the molecular geometry in solution in $CDCl_3$ appears,
from the n.m.r. spectra[21], to be of much lower symmetry than that of
the square antiprism or of the triangular-faced dodecahedron. The
thorium and uranium complexes, $M(DBM)_4$, have this latter geometry
in the solid state (the crystals are of orthorhombic symmetry[22], space
group *Pccn*), but in $CDCl_3$ solution the n.m.r. spectra indicate that the
structure is an undistorted square antiprism[13].

Nearly all of the recorded actinide(IV) compounds sublime
readily in a vacuum and this property provides a convenient way of
purifying them. They are insoluble, or only slightly soluble, in water or
diethyl ether, but are readily soluble in benzene, toluene, xylene or
chloroform, solvents which are therefore commonly used to extract
these compounds from aqueous media. The melting points of the
uranium(IV) complexes decrease as the length of the chain in R, for
the ligand CH_3COCH_2COR, increases, only the short chain or cyclic
derivatives (R = CH_3, C_2H_5, $CH_2OC_2H_5$, C_4H_3O, C_6H_5) being solid at
room temperature. The melting points of the corresponding complexes
derived from CF_3COCH_2COR show a similar trend, but are appreciably
higher, and only the longer-chain compounds (R = nC_4H_9, nC_5H_{11}) are
liquid at room temperature[3].

1:1 Adducts of $Th(AA)_4$ with ammonia and with aniline have
been recorded[12], and it is probable that other adducts of the less
sterically hindered β-diketonates could be prepared.

(c) Actinide(V) complexes

Pentakis diketone complexes are unknown, but uranium(V)
alkoxycompounds (p.201) have been reported, and a single protactinium
compound, the bright-yellow $Pa(AA)_2Cl_3$, has been recorded[23]. This is
prepared by treating the pentachloride with an excess of (H)AA in
dichloromethane.

(d) Actinide(VI) complexes

The uranyl(VI) complexes, UO_2L_2, are yellow, or orange to red, solids which, in most cases, are prepared simply by adding the ligand to aqueous methanolic or ethanolic solutions of uranyl nitrate[2,24] or acetate[25,26]. In a few instances it is necessary to increase the pH of the solution by adding alkali[2], but ammonia, primary, or secondary amines should not be used for this purpose because they often lead to the formation of β-ketoamine adducts; for example $[UO_2(AA)_2 \cdot CH_3COCHC\text{-}(NH_2)CH_3]$ is formed when ammonia is used[27]. The supposedly[28] anionic compounds of the type $R^+[UO_2(AA)_3]^-$, where R is NH_4, CH_3NH_3 or $C_6H_5NH_3$, are probably adducts of this type[27]. Uranyl(VI) complexes derived from the more sterically hindered β-diketones, such as (H)DBM, do, however, form simple 1:1 adducts with primary aliphatic amines[27] and a number of 1:1 adducts with ammonia, diethyl ether and pyridine are known[25,69]; analogous compounds with water, (H)AA, alcohols[29] and trimethylamine[27] are known for $UO_2(AA)_2$. The last, as well as $UO_2(TTA)_2$ and $UO_2(DBM)_2$, forms 1:1 adducts with diphenyl and dibenzyl sulphoxides[30], and $UO_2(TTA)_2$ forms similar adducts with tri-*n*-butyl phosphate and tri-*n*-octylphosphine oxide[31]. The uranium atom in these compounds is presumably 7-coordinate.

Very few physical measurements have been recorded for any of these compounds; the acetylacetonate, $UO_2(AA)_2$, is dimeric in benzene[28] and heat of solution measurements for a variety of the adducts formed with $UO_2(AA)_2$ suggest that the nitrogen-donor ligands are more strongly bonded to the uranium atom than are the oxygen donor ones[29].

3. TROPOLONE COMPLEXES

Tropolone ((H)T) can be regarded as the enolic form of a cyclic α-diketone; it forms *tetrakis* complexes, MT_4 (M = Th, pale yellow[32]; Pa, blue-green[33]; U, orange-brown[32]; Np, reddish-brown[33]; Pu, red-brown[34]) but, unlike (H)AA, it also forms anionic *pentakis* chelates,

$M^IM^{IV}T_5$, with the actinides(IV) of larger ionic radius (Th, Pa, U), where M^I is one of the smaller alkali metal cations. A yellow protactinium(V) complex, $PaT_4X(X = Cl, Br^{33})$ and the reddish-orange uranyl(VI)[35] and brown plutonyl(VI)[34] compounds have also been recorded.

The *tetrakis* complexes are prepared by adding the free ligand to an actinide tetrahalide in chloroform (Th[36]) or ethanol (Th, U, Np[33]), to the tetraacetate in methyl cyanide (U[32]) or to the hydrated tetra-nitrate in ethanol (Th[32]) or methanol (Pu[34]). They are precipitated (Th, Pa, U, Np) when lithium tropolonate is added to a solution of the actinide(IV) in oxygen free 1 M perchloric acid; if oxygen is present in the preparation of PaT_4 by this procedure, PaT_4ClO_4 is formed[33].

The uranium compound is moderately volatile and sublimes at $340°$ in a vacuum[32]. Although the crystal structures have not been determined, it is known that ThT_4 is dimorphic and that neither form is isostructural with any of the other actinide(IV) compounds. Infrared spectra and, in some instances, n.m.r. spectra are available in the references given for the preparation of these compounds.

The anionic complexes, $M^IM^{IV}T_5(M^{IV}Th; M^I = Li, Na, K)(M^{IV}Pa,$ U; $M^I = Li)$ are prepared by slurrying the *tetrakis* complexes with the alkali metal tropolonate in a mixture of methyl cyanide, water and ethanol (Th[32]) or methanol (U[32]) and by reaction with LiT in oxygen free N,N-dimethylformamide (DMF) (Th, Pa, U[33]). The isomeric isopropyltropolones form similar complexes to those obtained with tropolone itself (Th[32, 37], U[37]). The formation of these *pentakis* anions by the larger actinides(IV) is paralleled by the behaviour of the *tetrakis* complexes with oxygen-donor ligands. For example, 1:1 adducts with DMF are known[33] for ThT_4, PaT_4 and UT_4, but are not formed by NpT_4 or PuT_4. This last can be recrystallised from DMF without change[34]. Thorium[32] and uranium[33] complexes with dimethyl sulphoxide, $MT_4 \cdot DMSO$, are also known and it is probable that the analogous protactinium(IV) complex could be prepared.

Crystals of the thorium complex, $ThT_4 \cdot DMF$, are of triclinic symmetry (space group $P\bar{1}$) and the thorium atom is 9-coordinate[38]; the geometry of the molecule is a singly capped square antiprism in which one of the tropolonate groups spans a slant edge of the pyramidal cap on the square antiprism, and the DMF molecule is situated at another vertex of the cap such that the overall symmetry of the molecule is C_s (Fig. 25).

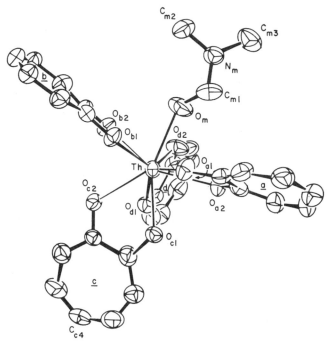

Fig. 25. Model in perspective of the $[ThT_4(DMF)]$ molecule (V.W. Day and J.L. Hoard, *J. Am. Chem. Soc.*, 92 (1970) 3626). Copyright (1970) by the American Chemical Society and reprinted by permission of the copyright holder.

The protactinium(V) complexes, PaT_4X (X = Cl, Br) are obtained by reaction of the pentahalide, or of the oxotribromide, with an excess of the ligand in dichloromethane. PaT_4Cl forms a dark-red 1:1 adduct with dimethyl sulphoxide and when treated with LiT in ethanol at 60° yields the yellow ethoxide, $PaT_4(OC_2H_5)$. This last reacts with 4 M perchloric acid to form the perchlorate PaT_4ClO_4, mentioned above. The perchlorate reverts to the chloride when treated with 6 M hydrochloric acid in acetone[33].

The actinide(VI) compounds, UO_2T_2 and PuO_2T_2, precipitate when the stoicheiometric quantity of tropolone is added to a uranyl(VI) salt[35] or to plutonyl(VI) nitrate[34] in methanol. The 1:1 adducts, $MO_2T_2 \cdot HT$, are formed when an excess of the ligand is used. All four compounds are soluble in DMF and in DMSO without, apparently, any evidence of complex formation. $UO_2T_2 \cdot HT$ desolvates at 155° and decomposes to uranium dioxide above 400°.

4. 8-HYDROXYQUINOLINE COMPLEXES

This ligand, commonly known as oxine, and the halogen or alkyl substituted ligand, are often used for the gravimetric estimation of thorium(IV) and uranium(VI). The complexes formed by the ligand and by some of its substitution products are listed in Table 30.

TABLE 30

ACTINIDE COMPLEXES OF 8-HYDROXYQUINOLINE AND SOME OF ITS SUBSTITUTION PRODUCTS

Complexes	Ligands[a] Hox	Hmox	Hcox	Hdcox
ML_3	M = Pu, Am[39]	–	Am[39]	Am[39]
ML_4^b	Th[40], U[41], Np, Pu[42]	Th[43], Np[42]	Np, Pu[42]	Th[43], U[41], Np, Pu[42]
MO_2L	Np(2 H_2O)[44] Pu[f]	Np, Pu[c]	Pu[d]	Pu[e]
MO_2L_2	U[43]	U[43]	–	U[43]

[a] Abbreviations: Hox = 8-hydroxyquinoline; Hmox = 2-methyl-Hox;
Hcox = 5-chloro-Hox; Hdcox = 5,7-dichloro-Hox.
[b] $ThL_4 \cdot HL$ and $UL_4 \cdot HL$ are well known, but the existence of UL_4 is less certain.
[c] $MO_2mox(Hmox) \cdot H_2O$[44].
[d] $PuO_2cox(Hcox)_2 \cdot H_2O$[44].
[e] $PuO_2dcox(Hdcox)_2 \cdot 2 H_2O$[44].
[f] $PuO_2ox(Hox)_2 \cdot xH_2O$[44].

(a) Actinide(III) complexes

Tris complexes of the type ML_3 are unknown for uranium(III) or neptunium(III), both of which are too readily oxidised to the quadrivalent state for it to be practicable to prepare them. Even the preparation of the yellow-green plutonium(III) complex, $Pu(ox)_3$, which is precipitated when solutions of the ligand and plutonium(III) are mixed, requires the presence of a holding reductant (sulphite and dithionite) in the solution and complete exclusion of oxygen, for the compound is very rapidly oxidised in air. These problems do not arise with the americium(III) complexes, however, and yellow-green $Am(ox)_3$, dark-green $Am(cox)_3$ and green $Am(dcox)_3$ are easily prepared by precipitation from aqueous solution at pH 5.5–6.5 (Hox) or from aqueous dioxane at about pH 6 (Hcox, Hdcox)[39].

The complexes formed by the halooxinates, $Am(cox)_3$ and $Am(dcox)_3$, are much more soluble in organic solvents than is $Am(ox)_3$; the u.v./visible spectra of these compounds have been reported[39], but nothing else is known about them. Although no other actinide(III) tris complexes have been recorded, solvent-extraction studies, with Hdcox as the extractant, indicate that the californium complex, $Cf(dcox)_3$, is the extracted species[46].

(b) Actinide(IV) complexes

Adducts of the type $ML_4 \cdot HL$ (M = Th[40,43], yellow to orange-red; U[41], brownish-yellow to gold) are precipitated from aqueous solutions of the actinide(IV) by Hox, or its derivatives, when the pH of the solutions is between 5 and 8, depending on the ligand. The extra molecule of ligand in $Th(ox)_4 \cdot Hox$ appears to be coordinated to the thorium atom by way of the phenolic oxygen atom, the acidic proton being on the nitrogen atom of the resulting monodentate ligand[47,48]. The thorium adducts yield the unsolvated complexes, ML_4, when heated in a vacuum[43], but, although free ligand sublimes from the uranium(IV) adduct under similar conditions[41], the unsolvated uranium(IV) complexes, UL_4, have not been identified with certainty. However, solvates are not formed by the higher actinides (*cf.* tropolone) and the yellow neptunium(IV) and red-brown plutonium(IV) complexes, $M(ox)_4$, are precipitated[42] from aqueous solution at pH 4.5–4.9 on the addition of Hox; the analogous Hcox and Hdcox complexes are prepared from aqueous dioxane (Np, Pu) because of the low solubility of the ligand in water. Partially hydrolysed species result when Hmox is added to aqueous solutions of neptunium(IV) or plutonium(IV), but $Np(mox)_4$ is obtained when the basic neptunium(IV) product is treated with hot, saturated aqueous Hmox, any excess of the ligand being removed from the product by vacuum sublimation[42]. The 5-7-dibromo- and 5-chloro-7-iodo-8-hydroxyquinoline complexes of neptunium(IV) and plutonium(IV) are made in the same way as $M(cox)_4$; they are brownish-yellow (Np) and dark-brown (Pu)[42].

The thorium complex, $Th(ox)_4 \cdot Hox$, yields a red solution when it is dissolved in a concentrated solution of Hox in dichloromethane; it has been suggested that the ion pair, $H_2ox^+[Th(ox)_5]^-$, may be present, and it is possible that the ligands in the anion complex, if the complex is formed, could all be bidentate[48]; this is a further analogy with the behaviour of tropolone. The thorium adduct, $Th(ox)_4 \cdot Hox$, also reacts with dimethyl sulphoxide, forming the orange complex[47] $Th(ox)_4 \cdot$ 2 DMSO, in which the thorium atom is also potentially 10-coordinate, and $Th(ox)_4 \cdot C_2H_5OH$ is formed in ethanol[48]. However, the additional molecule of ligand is not displaced by basic ligands, and with ethylenediamine (en) a complex of composition $(Th(ox_4)Hox)_2en$ is obtained, which may be a pentaoxinatocomplex[48], $(enH_2)^{2+}[Th(ox)_5]_2^-$.

Many other substituted 8-hydroxyquinoline complexes of thorium(IV)[43,48] and uranium(IV)[41] are known; these compounds are very similar in chemical behaviour to those discussed above and are not of very great importance.

(c) Actinide(V) complexes

The addition of the ligand to a solution of the actinide(V) in 0.1 M perchlorate solution at pH 6.5 (Hox) or pH 9 (Hmox) precipitates the yellow, hydrated neptunium(V) complexes, $NpO_2(ox) \cdot 2 H_2O$ and

$NpO_2(mox) \cdot H_2O$; the analogous plutonium(V) compounds, $PuO_2(ox)$-$(Hox)_2 \cdot xH_2O$ and $PuO_2(mox)(Hmox) \cdot H_2O$, precipitated under similar conditions, are green-brown and green respectively. The green complexes $PuO_2(cox)(Hcox)_2 \cdot H_2O$ and $PuO_2(dcox)(Hdcox)_2 \cdot 2\ H_2O$ are obtained[44] in the same way from aqueous dioxane at pH 5-6. Neptunium(V) also forms the complex anion, $[NpO_2(ox)_2]^-$, with an excess of the ligand; the yellow-green, monohydrated tetraphenylarsonium salt is precipitated[44] at pH 10.3-10.4. Very little is known about these compounds.

(d) Actinide(VI) complexes

The only actinide(VI) compounds known are the red uranyl(VI) complexes, $UO_2L_2 \cdot HL$; attempts[44] to prepare the neptunyl(VI) and plutonyl(VI) analogues failed, because the ligand reduced both sexivalent elements to lower oxidation states. The uranyl(VI) compounds are precipitated when the ligand is added to solutions of uranyl salts at pH 5-9; the actual value depends on the ligand used. Because of steric repulsions, which make it difficult for the relatively large HL molecule to approach the uranium atom[49], the 7-substituted ligand yields solvates of the type $UO_2L_2 \cdot X$, and not $UO_2L_2 \cdot HL$, where X is a smaller molecule (H_2O, $(CH_3)_2CO$, NH_3) present in the solution. The unsolvated *bis* chelates, UO_2L_2, result when the adducts, $UO_2L_2 \cdot HL$, are heated in a vacuum[43]. The oxinate so obtained, $UO_2(ox)_2$, appears to disproportionate in ethanol[52], forming the ethoxide, $UO_2(ox)(OC_2H_5)$, and $UO_2(ox)_2 \cdot Hox$. A number of substituted 8-hydroxyquinoline complexes of the type $UO_2L_2 \cdot HL$ additional to those listed in Table 30 are also known[43,49].

Crystals of $UO_2(ox)_2 \cdot Hox$ are of monoclinic symmetry (space group $P2_1/n$); the two ox groups are both bidentate, but the Hox molecule is unidentate, being bonded to the uranium atom by way of the phenolic oxygen atom only, so that the uranium atom is actually 7-coordinate. The geometrical arrangement is approximately a pentagonal bipyramid[50]. A molecule of chloroform, introduced during the recrystallisation of the compound from that solvent, was also found to be present in the lattice of the crystals used for the structure determination.

The unidentate molecule of Hox is displaced from $UO_2(ox)_2 \cdot$ Hox by oxygen-donor ligands, dark-red $UO_2(ox)_2 \cdot DMSO$ being obtained with dimethyl sulphoxide. The analogous urea complex, $UO_2(ox)_2 \cdot CO(NH_2)_2$, is prepared by adding Hox, followed by urea, to a solution of uranyl nitrate in 50% aqueous dioxane[51].

Orange or red anionic complexes, $M^I UO_2(ox)_3 (M^I = Na, (C_2H_5)_4N, (C_6H_5)_4As)$ are precipitated[53] when a solution of Hox in aqueous sodium carbonate is added to aqueous uranyl nitrate (Na salt) and when the alkylammonium bromide or phenylarsonium chloride is added to a solution of uranyl nitrate in a mixture of sodium hydroxide and Hox. The uranium atom is presumably 8-coordinate in the complex anion.

5. CUPFERRON N-NITROSO-N-PHENYLHYDROXYLAMINE) COMPLEXES

Very few cupferron complexes have been actually isolated and the compositions reported for those that have been obtained as solids usually depend solely on the determination of the metal content. The ligand, and other N-arylhydroxylamine compounds, are commonly used in the separation of the actinides by solvent extraction and, in some instances, for the gravimetric determination of the actinide element. In this last application, the precipitated metal complex is ignited to the metal oxide for weighing.

Uranium(III) cupferrate has not been isolated, but polarographic studies have been interpreted as indicating that this complex is formed in the reduction of uranyl cupferrate. It is said[54] to be soluble in ether and to be very stable, presumably with respect to oxidation. This is in marked contrast to the reported[55] failure of the attempt to prepare the plutonium(III) compound. The only result was the formation of the plutonium(IV) complex, and the reaction whereby the oxidation occurred requires further investigation.

The *tetrakis* complexes, $ML_4 (M = Th, white[56]; U, brown[57,58]; Pu, orange-yellow[55])$ are precipitated when an excess of the ligand is added to a dilute acid solution of the actinide(IV). The thorium complex is soluble in chloroform and in methylisobutyl ketone[59] and the uranium one is soluble in diethyl ether and in chloroform; the last is used to recrystallise the compound[57]

The uranyl complex of cupferron, UO_2L_2, does not appear to have been isolated; it is not precipitated from acid solution, but a chloroform extractable species, probably of this composition, may be present to a small extent in acid media[60]. Orange-yellow to orange complexes of N-acetyl- and N-nitrobenzoyl-N-phenylhydroxylamine, apparently of composition UO_2L_2, precipitate[61] from aqueous or ethanolic solutions of uranyl salts at pH 4.5–5.0 and it would be interesting to have structural data for these products. The attempted preparation of plutonyl(VI) cupferrate failed because of reduction to plutonium(IV) by the ligand[55].

The orange-yellow ammonium salt of the *tris* cupferrate anion, $[UO_2L_3]^-$, is obtained by mixing a solution of uranyl nitrate hexahydrate in methanol or ethanol with the ligand in 50% aqueous methanol. The alkali metal (Na, K) salts are also made by this procedure, whereas the rubidium and caesium salts are prepared by treating the sodium salt, the most soluble salt of this group, with a salt of the appropriate alkali metal in water or ethanol[62]. Crystals of the ammonium salt are of cubic symmetry (space group T^4-$P2_13$) but the molecular structure is unknown[62].

6. SCHIFF'S BASE COMPLEXES

The simplest ligand, salen, derived from salicylaldehyde and ethylenediamine, is reported to yield the uranium(IV) *bis* complex, U(salen)$_2$, when added to a solution of uranium tetrachloride in ethanol in the presence of sodium acetate[63]; the formation of an analogous thorium(IV) compound with the base derived from salicylaldehyde and 1-amino-8-naphthe 3,6-disulphonic acid is used for the colorimetric estimation of thorium[64]. The uranyl(VI) complexes, UO_2L_2, are prepared from uranyl acetate and the ligand in ethanol under reflux[71].

7. PHTHALOCYANINE (Pc, TETRABENZOPORPHYRAZINE) COMPLEXES

The dark-violet complexes $M(Pc)_2$ (M = Th[65], Pa[66], U[65]) are prepared by heating the actinide tetraiodide with phthalodinitrile at 240°.
The excess of the nitrile is extracted into benzene, and the products
are then sublimed from the residue at about $550°/10^{-5}$ torr. When
thorium tetrachloride is used instead of the tetraiodide in this preparation, a complex of the chlorinated phthalocyanine results[65]. There
is trace-scale evidence for the formation of $^{239}Np(Pc)_2$, prepared by
converting neutron-irradiated uranium metal to the tetraiodide and
treating this with phthalodinitrile in 1-chloronaphthalene[67]. It is possible
that $^{239}NpI_4$ is formed in the iodination of the irradiated metal, for
although neptunium tetraiodide is unknown, it may be stable with
respect to decomposition in a UI_4 matrix. Later work[70] has shown that
$^{237}Np(Pc)_2$, and even $Am(Pc)_2$, can be prepared from the triiodides by
this last reaction at 200°.

The phthalocyanine complexes are only slightly soluble in benzene
or in tetrahydrofuran, but they are quite soluble in pyridine,
1-chloronaphthalene and a few other organic solvents. Crystals of the
uranium complex are of monoclinic symmetry (space group $C2/c$) and
the uranium atom is 8-coordinate. The geometry with respect to the
nitrogen atoms is close to that of a square antiprism, but as the Pc ring
systems are rotated about 37° away from a prismatic configuration, the
deviation from the square antiprismatic geometry is significant. The Pc
groups are saucer-shaped, the four nitrogen atoms of each group being
at the base of the saucer[68] (Fig. 26).

The product which results from the reaction of uranyl chloride
with phthalodinitrile in boiling N,N-dimethylformamide appears to be
a complex of cyclopentakis (2 iminoisoindoline) and not a
phthalocyanine complex. It is probable that the molecule has a
pentagonal bipyramidal configuration with 5 nitrogen atoms in the
equatorial plane[70].

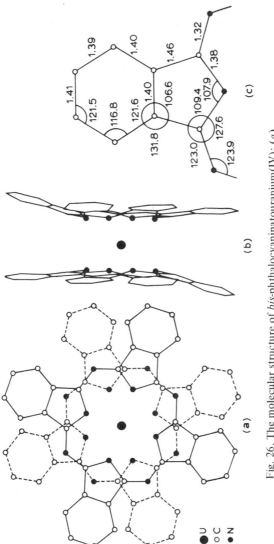

Fig. 26. The molecular structure of *bis*-phthalocyaninatouranium(IV): (*a*) Projection of the molecule onto the plane of the pyrrole nitrogen atoms; (*b*) Projection of the molecule along the crystallographic twofold axis. Only the pyrrole nitrogen atoms are marked. (*c*) Average bond lengths and angles in the Pc ring system (A. Gieren and W. Hoppe, *Chem. Commun.*, (1971) 413). Reprinted by permission of the Chemical Society, London.

8. REFERENCES

1   R.N. Haszeldine, W.K.R. Musgrave, F. Smith and L.M. Turton, *J. Chem. Soc.*, (1951) 609.
2   H.I. Schlesinger, H.C. Brown, J.J. Katz, S. Archer and R.A. Lad, *J. Am. Chem. Soc.*, 75 (1953) 2446.
3   H. Gilman, R.G. Jones, E. Bindschadler, D. Blume, G. Karmas, G.A. Martin Jr., J.F. Nobis, J.R. Thirtle, H.L. Yale and F.A. Yoeman, *J. Am. Chem. Soc.*, 78 (1956) 2790.
4   L. Wolf and H.J. Jahn, *Angew. Chem.*, 66 (1954) 754.
5   M. Ishidate and Y. Yamane, *Yokugaku Zasshi*, 77 (1957) 386, (*Chem. Abstr.*, 51 (1957) 11150).
6   C. Keller and H. Schreck, *J. Inorg. Nucl. Chem.*, 31 (1969) 1121.
7   M.D. Danford, J.H. Burns, C.E. Higgins, J.R. Stokeley Jr. and W.H. Baldwin, *Inorg. Chem.*, 9 (1970) 1953.
7a C.S. Erasmus and J.C.A. Boeyens, *Acta Cryst.*, A25 (1969) S162.
8   J.H. Burns and M.D. Danford, *Inorg. Chem.*, 8 (1969) 1780.
9   L.J. Nugent, J.L. Burnett, R.D. Baybarz, G.K. Warner, S.P. Tanner, J.R. Tanner and O.L. Keller Jr., *J. Phys. Chem.*, 73 (1969) 1540.
10 Y. Baskin and N.S. Krishna Prasad, *J. Inorg. Nucl. Chem.*, 25 (1963) 1011.
11 H.A. Swain Jr. and D.G. Karracker, *Inorg. Chem.*, 9 (1970) 1766.
12 W. Biltz, *Annalen*, 331 (1904) 334.
13 C. Wiedenheft, *Inorg. Chem.*, 8 (1969) 1174.
14 J.S. Dixon and C. Smith, in G.T. Seaborg, J.J. Katz and W.M. Manning (Eds.), *The Transuranium Elements*, Natl. Nucl. Energy Ser., Vol. IV–14B, McGraw-Hill, New York, 1949 p. 1108.
15 J.S. Dixon, J.J. Katz and E.F. Orlemann, *U.S. Patent* 2,989,556 (1944).
16 H. Titze and B. Allard, *Acta Chem. Scand.*, 24 (1970) 715.
17 D. Grdenić and B. Matković, *Acta Cryst.*, 12 (1959) 817.
18 D. Grdenić and B. Matković, *Nature*, 182 (1958) 465.
19 H. Titze, *Acta Chem. Scand.*, 24 (1970) 405.
20 A.E. Comyns, *Acta Cryst.*, 13 (1960) 278.
21 T.H. Siddall Tert. and W.E. Stewart, *Chem. Commun.*, (1969) 922.
22 V.L. Wolf and H. Bärnighausen, *Acta Cryst.*, 10 (1957) 605; 13 (1960) 778.
23 D. Brown and C.E.F. Rickard, 3rd Intern. Protactinium Conf., Schloss Elmau, Germany (1969); ZAED Rept. BMBW-FB K71-17, p. 21-1 (1971).
24 K.M. Abubacker and N.S. Krishna Prasad, *J. Inorg. Nucl. Chem.*, 16 (1961) 296.
25 L. Sacconi and G. Giannoni, *J. Chem. Soc.*, (1954) 2368, 2751.
26 A. Syamal, *J. Ind. Chem. Soc.*, 45 (1968) 904.
27 J.M. Haigh and D.A. Thornton, *J. Inorg. Nucl. Chem.*, 33 (1971) 1787.
28 J. Kager, *Z. Anorg. Allgem. Chem.*, 162 (1927) 82.
29 W.W. Wendlandt, J.L. Bear and G.R. Horton, *J. Phys. Chem.*, 64 (1960) 1289.
30 M.S. Subramanian and A. Viswanatha, *J. Inorg. Nucl. Chem.*, 31 (1969) 2575.

31   J.R. Ferraro and T.V. Healy, *J. Inorg. Nucl. Chem.*, 24 (1962) 1449, 1463.
32   E.L. Muetterties, *J. Am. Chem. Soc.*, 88 (1966) 305.
33   D. Brown and C.E.F. Rickard, *J. Chem. Soc. (A)*, (1970) 3373.
34   D.L. Plymale and W.H. Smith, *J. Inorg. Nucl. Chem.*, 31 (1969) 233.
35   D.L. Plymale and W.H. Smith, *J. Inorg. Nucl. Chem.*, 30 (1968) 2267.
36   E.L. Muetterties and C.M. Wright, *J. Am. Chem. Soc.*, 87 (1965) 4706.
37   E.L. Muetterties and C.W. Alegranti, *J. Am. Chem. Soc.*, 91 (1969) 4420.
38   V.W. Day and J.L. Hoard, *J. Am. Chem. Soc.*, 92 (1970) 3626.
39   C. Keller, S.H. Eberle and K. Mosdzelewski, *Radiochim. Acta*, 5 (1966) 185.
40   A. Corsini and J. Abraham, *Talanta*, 17 (1970) 439.
41   W. Wendlandt and G.R. Horton, *J. Inorg. Nucl. Chem.*, 19 (1961) 272.
42   C. Keller and S.H. Eberle, *Radiochim. Acta*, 4 (1965) 141.
43   G.R. Horton and W.W. Wendlandt, *J. Inorg. Nucl. Chem.*, 25 (1963) 241, 247.
44   C. Keller and S.H. Eberle, *Radiochim. Acta*, 8 (1967) 65.
45   T.J. la Chapelle, L.B. Magnusson and J.C. Hindman, in G.T. Seaborg, J.J. Katz and W.M. Manning (Eds.), *The Transuranium Elements*, Natl. Nucl. Energy Ser., Vol. IV-14B, McGraw-Hill, New York, 1949, p. 1108.
46   D. Feinauer and C. Keller, *Inorg. Nucl. Chem. Letters*, 5 (1969) 625.
47   A. Corsini and J. Abraham, *Chem. Commun.*, (1968) 856.
48   J. Abraham and A. Corsini, *Analyt. Chem.*, 42 (1970) 1528.
49   A. Corsini, J. Abraham and M. Thompson, *Chem. Commun.*, (1967) 1101.
50   D. Hall, A.D. Rae and T.N. Waters, *Acta Cryst.*, 22 (1967) 258.
51   A. Corsini and J. Abraham, *Canad. J. Chem.*, 48 (1970) 2360.
52   S. Oki, *Analyt. Chim. Acta*, 44 (1969) 315.
53   E.P. Bullwinkel and P. Nobel Jr., *J. Am. Chem. Soc.*, 80 (1958) 2955.
54   C.L. Rulfs and P.J. Elving, *J. Am. Chem. Soc.*, 77 (1955) 5502.
55   I.V. Moiseev, N.N. Borodina and V.T. Tsvetkova, *Russ. J. Inorg. Chem.*, 6 (1961) 277.
56   G.E.F. Lundell and H.B. Knowles, *Ind. Eng. Chem.*, 12 (1920) 344.
57   V. Auger, *Compt. Rend.*, 170 (1920) 995.
58   J.A. Holladay and T.R. Cunningham, *Trans. Am. Electrochem. Soc.*, 43 (1923) 329.
59   D. Dyrssen and V. Dahlberg, *Acta Chem. Scand.*, 7 (1953) 1186.
60   N.H. Furman, W.B. Mason and J.S. Pekola, *Analyt. Chem.*, 21 (1949) 1325.
61   N.N. Ghosh and G. Siddhanta, *J. Ind. Chem. Soc.*, 45 (1968) 1049; 46 (1969) 488.
62   W.S. Horton, *J. Am. Chem. Soc.*, 78 (1956) 897.
63   S.N. Poddar and D.K. Biswas, *Sci. Culture (Calcutta)*, 34 (1968) 117.
64   S.N. Poddar and K. Dey, *Ind. J. Chem.*, 3 (1965) 407.
65   F. Lux, D. Dempf and D. Graw, *Angew. Chem. Intern. Ed. Engl.*, 7 (1968) 819.
66   F. Lux, D. Brown, D. Dempf, R.D. Fischer and W. Hagenberg, *Angew. Chem. Intern. Ed. Engl.*, 8 (1969) 894.
67   F. Lux, F. Ammentorp-Schmidt, D. Dempf, D. Graw and W. Hagenberg, *Radiochim. Acta*, 14 (1970) 57.

68 A. Gieren and W. Hoppe, *Chem. Commun.*, (1971) 413.
69 J.M. Haigh and D.A. Thornton, *J. Mol. Struct.*, 8 (1971) 351.
70 W. Hagenberg, R. Gradl and F. Lux, GDCh–Hauptversammlung, GDCh–Fachgruppe "Kern-, Radio- und Strahlenchemie", Karlsruhe, September 1971.
71 P. Pfeiffer, T. Hesse, H. Pfitzner, W. Scholl and H. Thielert, *J. Prakt. Chem.*, 149 (1937) 217.

Chapter 11

BOROHYDRIDES, ORGANOMETALLIC COMPOUNDS AND AMIDES

1. INTRODUCTION

Actinide compounds in which the metal is bonded to boron, carbon or nitrogen, other than the simple borides, carbides or nitrides, have not been investigated to a great extent. This is largely because of the experimental difficulties of handling them, for nearly all are reactive towards moisture and atmospheric oxygen. Most of the published work described in this chapter refers to compounds of thorium and uranium; much less is known about the corresponding compounds of protactinium and the transuranium actinides.

2. BOROHYDRIDES

Thorium(IV)[1] and uranium(IV)[2] borohydrides, $M(BH_4)_4$, both of which are readily hydrolysed by water, are easily obtained by treating the actinide tetrafluoride with aluminium borohydride at room temperature:

$$MF_4 + 2 Al(BH_4)_3 \longrightarrow M(BH_4)_4 + 2 AlF_2(BH_4)$$

The white thorium compound is, incidentally, the most volatile compound of that element to be recorded. It melts, with decomposition, at 203° and sublimes at 150° in a vacuum, which provides a simple method of purification; the compound decomposes rapidly to thorium boride at 300°. It is very soluble in diethyl ether, forming a *bis* etherate, and is also soluble in tetrahydrofuran, but is insoluble in benzene. It is isomorphous[1] with the corresponding dark-green uranium compound, which is also rather volatile, but is much less stable to heat than the thorium compound, decomposing slowly at room temperature and more rapidly at 100° to yield the uranium(III) compound[2]:

$$2\ U(BH_4)_4\ \longrightarrow\ 2\ U(BH_4)_3\ +\ H_2\ +\ B_2H_6$$

Decomposition to uranium boride occurs at 150–200° and it is likely that the thermal stability of the transuranium actinide(IV) compounds, when prepared, will prove to be even less.

$U(BH_4)_4$ reacts with methanol to form the tetramethoxide and with hydrogen chloride to yield uranium tetrachloride[2]. With trimethylborane, $B(CH_3)_3$, at 50–70° a mixture of products is obtained which includes $U(BH_4)_3(CH_3BH_3)$ and $U(CH_3BH_3)_4$, both of which are volatile. Reactions of a similar nature occur with the higher trialkylboranes, but the products have not been identified[3].

3. ORGANOMETALLIC COMPOUNDS

Compounds in which carbon is σ-bonded to the actinide-metal atom have not been recorded, but π-bonded complexes with allyl, cyclopentadienyl and cyclooctatetraenyl groups are now well known. The first of these compounds, $U(C_5H_5)_3Cl$, was reported[4] as recently as 1956, but since that time preparative work has extended the range of actinide elements studied as far as californium. Three reviews[5-7], covering the literature to 1970, have now appeared and give comprehensive accounts of the properties of these compounds, including a detailed interpretation of their n.m.r., u.v./visible and infrared spectra. The review by Hayes and Thomas[6] is particularly useful in these respects.

(a) π-Allyl complexes

Tetraallylthorium (dark-yellow[8]) and uranium (dark-red[9]), $M(C_3H_5)_4$, are the least stable to heat of any of the known actinide organometallic compounds. The former can be kept without decomposition only below $0°$ and the latter only below $-20°$. Above $-20°$ the uranium compound decomposes with the liberation of a mixture of propane (18.5%) and propene (81.5%). It is spontaneously inflammable in air. The proton n.m.r. spectra[10] of the thorium and uranium compounds are typical of those that would be expected for π-bonded complexes.

The two compounds are prepared by treating the metal tetrahalide with an allyl Grignard reagent in ether at low temperature:

$$MX_4 + 4 C_3H_5MgY \longrightarrow M(C_3H_5)_4 + 4 MgXY$$

(b) Cyclopentadienyl complexes

Simple *tris* cyclopentadienyl complexes are known for uranium and for the elements from plutonium to californium inclusive (Table 31); the neptunium compound should also be obtainable, but does not seem to have been prepared. It is, however, unlikely that the thorium or protactinium *tris* complex will be preparable, particularly in view of the lack of conclusive evidence for the existence of compounds of these elements in their terpositive oxidation states. Berkelium provides the only example of a *bis* cyclopentadienyl actinide chloride to be recorded to date.

The *tetrakis* cyclopentadienyl complexes of the elements from thorium to neptunium (Table 32) give rise to the corresponding *tris* cyclopentadienyl actinide halides (Table 33) which are useful starting materials for the preparation of other *tris* cyclopentadienyl actinide derivatives. One example of a *bis* cyclopentadienyl actinide dihalide(U) has also been recorded.

The methods used for the preparation of the *tris* and *tetrakis* cyclopentadienyl compounds involve the reaction of the appropriate metal halide with an ionic cyclopentadienide, either in a non-aqueous medium such as benzene[11] or tetrahydrofuran (THF), the latter yielding the adduct $M(C_5H_5)_3 \cdot THF$ (method A):

$$UCl_3 + 3 KC_5H_5 \longrightarrow U(C_5H_5)_3 + 3 KCl$$

TABLE 31

ACTINIDE *tris* CYCLOPENTADIENYL COMPLEXES

$M(C_5H_5)_3$, M =	U	Pu	Am	Cm	Bk	Cf
References	11	12	13	15, 16	17	17
Colour	Bronze	Moss green	Rose (or[14] orange-yellow)	Colourless	Amber	Ruby-red
Sublimation temperature[a], °C	120–150(d)[b] 140–165		160–200	180	135–165	135–220
Preparative method (see text)	A,D	B	B	B(^{248}Cm), C(^{244}Cm)	B	B

[a] Temperatures refer to sublimation in a vacuum, usually about 10^{-5} torr.
[b] The greater part of the compound decomposes and only trace amounts sublime.

TABLE 32

ACTINIDE *tetrakis* CYCLOPENTADIENYL COMPLEXES

$M(C_5H_5)_4$,	$M = Th$	Pa	U	Np
References	18, 19	20	19, 21	22
Sublimation(s)[a] or decomposition(d) temperature, °C	170–290(s)	220(d)	200–220(d)	200–220(d)
Colour	Colourless	Orange	Red	Brown-red
Preparative method (see text)	A,C	B	A,C	A

[a]Sublimation temperatures refer to high vacuum.

TABLE 33

ACTINIDE *tris* CYCLOPENTADIENYLHALIDES[a], $M(C_5H_5)_3X$

M/X	F	Cl	Br	I
Th	Pale-yellow (23) sub.[b] 200°	Colourless (24) sub. 200°	Pale-yellow (23) sub. 180°	Pale-yellow (23) sub. 190°
U	Green (23,25) sub. 170°	Pale-brown (4) sub. 120–130°	Dark-brown (25) sub. 160°	Brown (25) sub. 170°
Np	Green (26,27) sub. 170°	Dark-brown (26) sub. 100–105°		

[a]References in parentheses.
[b]Sublimation temperatures refer to high vacuum.

or, in the cases of beryllium dicyclopentadienide[12,13,16,17] (method B) and the magnesium analogue[15] (method C), as melts without an added solvent:

$$2\,M^{III}Cl_3 \;+\; 3\,M^{II}(C_5H_5)_2 \;\longrightarrow\; 2\,M^{III}(C_5H_5)_3 \;+\; 3\,M^{II}Cl_2$$

For the preparation of the *tetrakis* complexes, a tetrachloride is used when following method A, a protactinium pentahalide (a tetrahalide would react in the same way) with method B and an actinide tetrafluoride with method C. The last is a particularly useful way of making the thorium and uranium *tetrakis* complexes[19]; in this instance, the reaction is best carried out with the reagents heated in a sealed tube at 180-280°.

Preparations of the *tris* complexes made by methods B and C are usually separated from the reaction mixture by vacuum sublimation; these two methods are to be preferred for the preparation of complexes formed by the more intensely radioactive elements because inflammable solvents are not involved, and the risk of an explosion in the glove-boxes used for work with such elements is eliminated.

Because of the low thermal stability of the *tetrakis* complexes, they are normally extracted from the residue into a hydrocarbon solvent, such as benzene, after the reaction mixture has been evaporated to dryness (preparation by method A) or from the reaction mixture itself after removing any excess of the dicyclopentadienide used in the preparation by vacuum sublimation (methods B and C). The *tetrakis* thorium complex can, however, be isolated by sublimation techniques.

In addition to these procedures, the uranium *tris* complex has been prepared by the reduction of the *tetrakis* complex with potassium in benzene (method D); finely divided, highly reactive uranium metal is formed and this brings about the reduction:

$$U + 3 U(C_5H_5)_4 \longrightarrow 4 U(C_5H_5)_3$$

(c) Tris cyclopentadienyl actinides

The actinide *tris* complexes are soluble in hydrocarbon solvents, such as benzene, and in tetrahydrofuran, with which they form adducts. The *tris* complexes are stable to heat, apart from the uranium compound which decomposes at 120-150°, and, with that exception, they sublime at moderate temperatures in a vacuum (Table 31). All of the complexes decompose on exposure to air, the uranium and plutonium compounds being pyrophoric, and all are hydrolysed by water with the liberation of cyclopentadiene and formation of the actinide hydroxide.

Crystals of the berkelium and californium *tris* cyclopentadienyls are of orthorhombic symmetry (space group *Pbcm*), isostructural[16, 17] with the samarium compound. The structure of the last consists of two separate, infinite zigzag chains of $Sm(C_5H_5)_3$ molecules; each samarium atom in these chains is surrounded by three cyclopentadiene rings, to which the metal atom is bonded, and a fourth ring from an adjacent unit is sufficiently close to make up a distorted tetrahedral array of cyclopentadiene rings about the metal atom (Fig. 27). The two chains

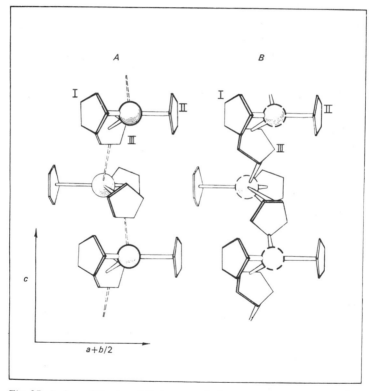

Fig. 27. A view of the A and B structures in $Sm(C_5H_5)_3$ along the [120] direction showing the difference between the two structures. A dashed circle for Sm(*B*) implies its diffuseness; dashed bonds indicate that the bonds are long (C.-H. Wong, T.-Y. Lee and Y.-T. Lee, *Acta Cryst.*, B25 (1969) 2580). Reproduced by permission of the International Union of Crystallography.

differ in that in one of them the three bonded cyclopentadiene rings are almost equidistant from the metal atom (Sm–ring distance in the samarium compound, 249–255 pm), which lies on the normal to the centre of each ring, whereas in the other chain one of the three bonded rings (B III, Fig. 27) is displaced by 60 pm from the normal to the centre of the ring and along the plane of the ring, as compared with the position observed for the corresponding cyclopentadiene ring (A III) in the first chain[28]. The metal–carbon bond in the samarium compound has appreciable ionic character, in agreement with its chemical behaviour, and that of the other lanthanide *tris* complexes, which yield ferrocene with ferrous chloride[4]. However, it is not clear whether the actinide *tris* complexes react in this way with ferrous chloride. Structural data are not available for the uranium, plutonium, americium and curium *tris* complexes, but it is probable that they also will be found to be isostructural with the samarium compound.

The lanthanide *tris* cyclopentadienyls behave as Lewis acids and the analogous uranium *tris* complex is now known[11] to exhibit the same behaviour; it forms 1:1 adducts with basic ligands such as isonitriles (cyclohexylisonitrile, red-brown adduct, slightly soluble in benzene) and 1-nicotine (black-brown adduct). In the latter, both of the nitrogen atoms of the ligand appear to be bonded to the uranium atom. *Tris* cyclopentadienyl plutonium also forms a 1:1 adduct with cyclohexylisonitrile[29]. In this last, the C–N band in the infrared spectrum, which appears at 2130 cm^{-1} in the free ligand, is shifted to 2190 cm^{-1} in the adduct, close to the frequencies observed for the lanthanide 1:1 adducts, whereas in the uranium adduct the band appears[11] at 2160 cm^{-1}. It has been suggested[11] that the M–C bond in the uranium compound may possess more marked σ-donor character than that in the corresponding lanthanide or plutonium adducts and that there may possibly be some back donation from the uranium metal orbitals to the ligand in the case of the uranium compound. An increase of the σ-donor character of the M–C bond would tend to promote the contribution of the form $\overset{-}{C} \equiv \overset{+}{N}:R$ of the ligand, so increasing the N–C bond order and hence raising its stretching frequency. On the other hand, a π contribution from the metal orbitals, if there is any, would

.tend to reduce the bond order[54]. One might expect that the degree of this back donation would decrease along the actinide series with increasing atomic number, reflecting the spatial contraction of the metal f-orbitals caused by the increase in effective nuclear charge to which they are exposed. The C–N stretching frequency would then increase until the π contribution became zero. It is not yet clear whether it is necessary to invoke the two effects to explain these changes and another alternative, namely that the σ-acceptor properties of the metal could change along the series, may also play some part. To clarify these points it is obviously desirable to have more information on these and other actinide adducts with isonitriles.

Tris cyclopentadienyl curium (prepared from the long-lived isotope ^{248}Cm) exhibits[16] a bright-red fluorescence when exposed to ultraviolet radiation ($\lambda = 360$ nm); the same compound, prepared from the short-lived, intensely α radioactive isotope, ^{244}Cm, exhibits the same fluorescence without exposure to ultraviolet radiation as a result of excitation by its own intense α-radiation[15].

Spectroscopic data (u.v./visible, infrared, ^{1}H-n.m.r.) and mass spectra are available in most of the original papers which describe the preparation of the complexes and their adducts. Summaries of these, and associated magnetic data, are given in the reviews[6, 7].

(d) Bis cyclopentadienyl berkelium chloride

When berkelium *tris* cyclopentadienyl is prepared by method B (p. 234), some *bis* cyclopentadienyl berkelium chloride is also formed; it can be separated from the *tris* complex by fractional sublimation, for it is somewhat less volatile than the latter, subliming between 220 and 300° in a vacuum. The compound is isostructural[30] with its samarium analogue, which is a dimer, $[(C_5H_5)_2SmCl]_2$, in which both chlorine atoms bridge the two metal atoms.

(e) Tetrakis cyclopentadienyl actinides

These compounds are moderately soluble in hydrocarbons, such as benzene or *n* pentane, in halogenated hydrocarbons, such as chloro-

form, dichloromethane or bromoform, and in tetrahydrofuran. They
are decomposed by water or mineral acids, and are moderately
sensitive to atmospheric oxygen. They are less stable to heat than the
tris complexes and are less volatile than the latter. Reduction to the
finely divided metals occurs when the complexes are treated with
potassium suspended in benzene[31] (*cf.* method D for the preparation of
$U(C_5H_5)_3$, p. 000). The *tetrakis* uranium complex reacts with halogens[31]
(Cl_2, Br_2, I_2) yielding the *tris* cyclopentadienyl uranium halide,
$(C_5H_5)_3UX$. The other known *tetrakis* complexes should react with
halogens in the same way.

No crystallographic data have been published for the *tetrakis* complexes; the infrared spectra of the four known compounds are, however,
almost identical and it is therefore probable that all have a similar
structure[32]. More information is available for the *tetrakis* uranium
complex; its dipole moment is zero[21], so that it is very probable that
the four cyclopentadiene rings are disposed about the uranium atom at
the apices of a tetrahedron enclosing the latter. Moreover, all of the
ring protons in each of the known complexes are equivalent, for only
one proton signal is observed in their ^{1}H-n.m.r. spectra.

(f) Tris cyclopentadienyl actinide halides

Nearly all of these compounds (Table 33) are prepared by the
general methods A, B and C described for the *tris* and *tetrakis*
cyclopentadienyl complexes (p. 232, 234), but rather less than the
stoicheiometric quantity of the ionic cyclopentadienide should be used
than would be required for the reactions below:

$$MX_4 + 3 M^I C_5H_5 \rightarrow (C_5H_5)_3MX + 3 M^I X$$

$$2 MX_4 + 3 M^{II}(C_5H_5)_2 \rightarrow 2(C_5H_5)_3MX + 3 M^{II}X_2$$

The two exceptions are *tris* cyclopentadienyl thorium chloride,
$(C_5H_5)_3ThCl$, and the analogous uranium fluoride, $(C_5H_5)_3UF$. The
thorium compound is precipitated when thorium tetrachloride is
treated with a large excess of potassium cyclopentadienide in diethyl

ether at room temperature[24]; the insolubility of the compound in this medium evidently prevents the reaction going to completion so as to yield the *tetrakis* complex. The analogous uranium fluoride has only been prepared by halogen exchange, the corresponding bromide being heated with sodium fluoride[23]. There is no reason to suppose that it would not be formed by treating uranium tetrafluoride with the appropriate quantity of an ionic cyclopentadienide, as in the case of the neptunium compound, but the reaction does not seem to have been attempted. The two known *tris* cyclopentadienyl neptunium halides are also formed when the neptunium trihalides are treated with molten beryllium dicyclopentadienide[26]; the mechanism whereby the oxidation occurs is not known.

All of the known *tris* cyclopentadienyl actinide halides sublime in a vacuum, providing a convenient way of purifying these compounds; they are appreciably more stable to heat than are the *tetrakis* complexes and are soluble in a range of organic solvents similar to that in which the latter dissolve. They are particularly sensitive to oxidation by atmospheric oxygen when in solution.

The metal–chlorine bond in *tris* cyclopentadienyl uranium chloride is ionic. Crystals of the compound are of monoclinic symmetry (space group $P2_1/n$) and the structure of the molecule consists of a tetrahedral arrangement of the three cyclopentadiene rings and the chlorine atom about the central uranium atom[33] (Fig. 28). The infrared and u.v./visible spectra of the uranium chloride, bromide and iodide are almost identical[25] and they probably all have the same molecular structure. However, the uranium fluoride complex is a fluorine-bridged dimer; its infrared spectrum has been recorded, and the uranium-fluorine stretching vibration appears[25] at 466 cm^{-1}. It is interesting to note that this fluoride complex forms 1:1 adducts with some of the *tris* cyclopentadienyl lanthanides, evidently by way of a bridging fluorine atom, for in these adducts the uranium-fluorine stretching vibration appears[34] at lower frequencies than those observed for the dimer, namely at 432 cm^{-1} in $(C_5H_5)_3UF \cdot Yb(C_5H_5)_3$ and 423 cm^{-1} in $(C_5H_5)_3UF \cdot U(C_5H_5)_3$.

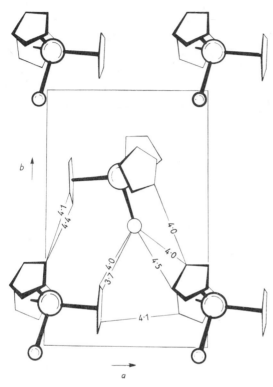

Fig. 28. The structure of $(C_5H_5)_3UCl$ viewed down the c axis (from 0 to ½ c) (C.-H. Wong, T.-U. Yen and T.-Y. Lee, *Acta Cryst.*, 18 (1965) 340). Reproduced by permission of the International Union of Crystallography.

Alkoxides of composition $(C_5H_5)_3MOR$, where M is thorium or uranium and R is CH_3, C_2H_5, nC_4H_9 and tC_4H_9, are formed by treating the appropriate metal *tris* cyclopentadienyl halide with the alkali metal alkoxide in a non-aqueous solvent. Other *tris* cyclopentadienyl uranium alkoxides of this type, where R = C_3H_7, nC_6H_{13}, nC_8H_{17} and *cyclo* C_6H_{11}, are made in the same way. This preparative method has also been applied to the preparation of *tris* cyclopentadienyl uranium thiocyanate, cyanate, borohydride and hydroxide, but the corresponding uranium tetrafluoroborate, $(C_5H_5)_3UBF_4$, is made by treating the *tris* cyclopentadienyl fluoride with boron trifluoride in benzene. A more detailed description of these compounds has been given in the recent reviews[5-7].

An investigation[35] of the mass spectra of the *tris* cyclopentadienyl actinide halides has shown that the nature of the principal fragments formed from these compounds depends to a great extent on the strength of the metal–halogen bond. Where this bond is strong, as in the fluoride, a cyclopentadiene ring is lost (*e.g.* $(C_5H_5)_3ThF$ forms mainly the fragment $(C_5H_5)_2FTh^+$) and where it is weak, as in the iodide, the halogen is lost (*e.g.* $(C_5H_5)_3ThI$ forms the fragment $(C_5H_5)_3Th^+$).

(*g*) *Bis cyclopentadienyl uranium dichloride*

The greenish-brown solid dichloride, $(C_5H_5)_2UCl_2$, is prepared by treating uranium tetrachloride with the stoicheiometric quantity of thallous cyclopentadienide, TlC_5H_5, in dimethoxyethane. It does not sublime when heated, but decomposes[36].

(*h*) *Tris cyclopentadienyl uranium dialkoxides and diphenoxide*

Green solids of composition $(C_5H_5)_3U(OR)_2$, where R is CH_3, nC_4H_9 and C_6H_5, are reported[37] to be formed by the reaction of the bromodichlorodialkoxide (or phenoxide), $UCl_2Br(OR)_2$ (p. 202, 205), with sodium cyclopentadienide in tetrahydrofuran or diethyl ether under reflux:

$$UCl_2Br(OR)_2 + 3\,NaC_5H_5 \longrightarrow (C_5H_5)_3U(OR)_2 + 2\,NaCl + NaBr$$

These compounds are rather volatile; the butoxide and phenoxide sublime at $125-150°/10^{-3}$ torr and the methoxide at $140°/10^{-4}$ torr. Further information has not been published concerning these species, but it is quite possible that they are actually alcohol or phenol adducts of the *tris* cyclopentadienyl alkoxide or phenoxide.

(*i*) *Bis cyclooctatetraenyl actinides*

Compounds of the type $(C_8H_8)_2M$ (M = Th, bright-yellow[38]; U, green[39]; Np[40], Pu[40]) are prepared by treating the actinide tetrachloride (Th^{38}, U^{39}, Np^{40}), or a hexachlorocomplex of the type $A_2^I MCl_6$ (U, Pu)[40] with potassium cyclooctatetraenide in dry, oxygen free tetrahydrofuran at temperatures below $0°$. Apart from the thorium complex, which is decomposed by water and is isolated from the reaction mixture by evaporating the solvent, they are recovered from the tetrahydrofuran solution in which they have been formed by precipitation with water, followed by extraction of the precipitate into benzene or toluene. All four of the complexes decompose on exposure to air; the uranium one is particularly reactive in this respect and inflames in air. The thorium compound sublimes at $160°/0.01$ torr and decomposes above $190°$; it explodes at red heat. The uranium compound is also rather volatile, subliming at $180°/0.03$ torr, but similar information for the neptunium and plutonium compounds is lacking. Although the colours of the last two have not been recorded, it is known that a solution of the neptunium compound in toluene is yellow in thin layers and blood-red in thick layers, whereas a solution of the plutonium compound in that solvent is cherry-red. Apart from the thorium compound, which is insoluble in most organic solvents, they are sparingly soluble in hydrocarbons, such as chloroform and carbon tetrachloride. The thorium compound does, however, dissolve in dimethyl sulphoxide, but the n.m.r. spectrum of the solution exhibits a complex multiplet at 6.2 p.p.m. in place of the sharp singlet to be expected for a D_{8h} sandwich structure; thus it seems probable that a complex is formed with this solvent.

Crystals of all four compounds are of monoclinic symmetry (space group $P2_1/n$), and the molecular structure of the uranium compound has been shown[41] to be a sandwich with the C_8H_8 rings eclipsed, the overall symmetry being D_{8h}. The higher actinides may well form sandwich complexes of the type $KM(C_8H_8)_2$, such as those formed[42] by the terpositive lanthanides.

The intensities of the absorption bands in the visible spectra of the uranium, neptunium and plutonium compounds are at least an order

of magnitude greater than the intensities normally observed for f-f
transitions in other complexes of these actinide elements, an
enhancement in intensity which seems to indicate an unusual degree of
mixing between the $5f$ and $6d$ orbitals[40]. There is also a large, positive
isomer shift ($+2.5$ cm sec^{-1}) for neptunium(IV) in the Mössbauer
spectrum of $Np(C_8H_8)_2$; this indicates a considerably increased shielding
of the $6s$ shell, which in turn suggests a large electron contribution from
ligand to metal orbitals[40].

4. AMIDES AND IMIDES

Thorium tetraamide, $Th(NH_2)_4$, is reported[43] to be formed when
the tetrachloride ammine, $ThCl_4 \cdot 4 NH_3$, is heated at 250-300° in dry
ammonia. It is said to decompose to the di-imide, $Th(NH)_2$, at a dull
red heat, and this, in turn, apparently yields the nitride at bright red
heat in an atmosphere of either hydrogen or ammonia. More recently, it
has been reported that the white imidediamide, $Th(NH_2)_2NH$, is
precipitated when potassium amide is added to a solution of the
hexanitratothorate(IV), $K_2Th(NO_3)_6$, in anhydrous liquid ammonia at
$-10°$. With an excess of potassium amide the products are, apparently,
amidoimide complexes, their compositions being $KTh(NH_2)(NH)_2$,
$KTh(NH_2)_3(NH)$ and $K_2Th_2(NH_2)_4(NH)_3$, depending on the quantity
of potassium amide that is used[44]. They are white pyrophoric solids,
which are readily hydrolysed; the last two of the compounds
decompose at 20°, but are apparently quite stable at 0°. The
imidediamide decomposes at 50° to yield a product of composition
$Th_2(NH)_3(NH_2)_2$ and at 100° the diimide, $Th(NH)_2$, appears to be formed. Above 130° the last compound is reported[44] to yield the nitride,
Th_3N_4. This is in marked contrast to the behaviour reported for the
tetraamide and di-imide noted above; indeed these results cast
considerable doubt on the original findings. It would obviously be
useful to have more information on these systems.

The reaction of uranium tetrachloride with potassium amide in
anhydrous liquid ammonia appears to be rather less complex; with a
large excess of potassium amide ($UCl_4 : KNH_2 = 1:20$) the product is

$K_4U(NH)_4$, whereas with a smaller excess (UCl_4: KNH_2 = 1:7) the product is $K_2U(NH)_3$. In contrast to this, sodium amide yields only $U(NH)_2$ under similar conditions[45]. These products may be formed by way of the intermediate $U(NH_2)_4$, but this compound has not been isolated.

The analogous reaction of potassium dioxotrinitratouranate(VI) with potassium amide in liquid ammonia yields a brown precipitate of uranyl amide, $UO_2(NH_2)_2$, which decomposes at $148°$ in gaseous ammonia under pressure to yield the brown imide. However, an imidocomplex, $K_2UO_2(NH)_2$, results when the preparation of the amide is carried out with an excess of potassium amide[46]. It would be useful to have structural information about these compounds, for the compounds described as amides could well be ammonia adducts of imides. For example, $UO_2(NH_2)_2$ may really be $UO_2(NH)\cdot NH_3$.

Partially ammonolysed species derived from the halides are also known. The reaction of thorium tetraiodide with potassium amide in liquid ammonia yields[47] the diamidospecies, $Th(NH_2)_2I_2\cdot 3\ NH_3$, and uranium trichloride[48] reacts with ammonia at $450-500°$ to form the amidochlorides $U(NH_2)Cl_2$ and $U(NH_2)_2Cl$. These yield the imidochloride, $U(NH)Cl$, and nitrides at progressively higher temperatures. The imidochloride crystallises in the tetragonal PbFCl structure (space group $P4/nmm$-D_{4h}^7). The other uranium(III) imidohalides, $U(NH)X$ (X = Br, I), are thought to be formed as intermediates in the high-temperature ammonolysis of the appropriate trihalides[50].

Uranium tetrachloride yields[48] the amidotrichloride, $U(NH_2)Cl_3$, at $450°$ in ammonia and this decomposes to uranium trichloride at $475°$. The amidotrichloride, a grey-green solid, is also formed when the octammine, $UCl_4\cdot 8\ NH_3$, is heated at $240-250°$ in a vacuum[49].

Tetra(diethylamides), $M[N(C_2H_5)_2]_4$, are known for thorium and uranium; they are prepared by treating the actinide tetrachloride with lithium diethylamide in organic solvents such as diethylamine, ether, n-hexane or tetrahydrofuran, followed by vacuum sublimation of the product from the residue left when the solvent is evaporated. The yield in the case of the yellow thorium compound is very low[51]; the sublimation occurs at $130-140°/0.1$ torr, but at these temperatures the losses are very considerable owing to thermal decomposition. In fact, the preparative yield is close to theoretical in the cold[52], but the compound is extremely difficult to purify in any other way but by vacuum sublimation. The metal–nitrogen stretching mode in the infrared spectrum of the compound appears[51] at 540 cm^{-1}.

The yield of the corresponding uranium compound by the vacuum sublimation technique[53] is 25–30% of theoretical; it is also formed when uranium pentachloride is treated with lithium diethylamide. The corresponding uranium dimethyl, diisopropyl, di-n-butyl and diallyl derivatives are probably formed in the reaction of uranium tetrachloride with the appropriate lithium dialkylamide, but they could not be isolated from the reaction mixture by vacuum sublimation because of their thermal decomposition[53].

These compounds are very sensitive to the action of oxygen and water, but dissolve in non-polar solvents, such as benzene, diethyl ether, and petroleum ether without decomposition. The uranium compounds are known to react with alcohols to form alkoxides (p. 199) and with mercaptans to form mercaptides (p. 205); the thorium compounds presumably react in the same manner. Carbon dioxide, disulphide and diselenide insert into the metal–nitrogen bond of the diethylamides to form the corresponding carbamates[52], but little else is known about the behaviour of these compounds.

5. REFERENCES

1 H.R. Hoekstra and J.J. Katz, *J. Am. Chem. Soc.*, 71 (1949) 2488.
2 H.I. Schlesinger and H.C. Brown, *J. Am. Chem. Soc.*, 75 (1953) 219.
3 H.I. Schlesinger, H.C. Brown, L. Horvitz, A.C. Bond, L.D. Tuck and A.O. Walker, *J. Am. Chem. Soc.*, 75 (1953) 222.
4 L.T. Reynolds and G. Wilkinson, *J. Inorg. Nucl. Chem.*, 2 (1956) 246.
5 H. Gysling and M. Tsutsui, *Advan. Organomet. Chem.*, 9 (1970) 361.
6 R.G. Hayes and J.L. Thomas, *Organomet. Chem. Rev.*, 7 (1971) 1.
7 B. Kanellakopulos and K.W. Bagnall, *Intern. Rev. Sci.*, 7 (1972) 299; Med. Techn. Publ., Aylesbury and Butterworth, London.
8 G. Wilke, B. Bogdanovic, P. Hardt, P. Heimbach, W. Keim, M. Körner, W. Oberkirch, K. Tanaka, E. Steinrücke, D. Walter and H. Zimmerman, *Angew. Chem.*, 78 (1966) 157.

9 G. Lugli, W. Marconi, A. Mazzei, N. Paladino and U. Pedretti, *Inorg. Chim. Acta*, 3 (1969) 253.

10 N. Paladino, G. Lugli, U. Pedretti and M. Brunelli, *Chem. Phys. Letters*, 5 (1970) 15.

11 B. Kanellakopulos, E.O. Fischer, E. Dornberger and F. Baumgärtner, *J. Organomet. Chem.*, 24 (1970) 507.

12 F. Baumgärtner, E.O. Fischer, B. Kanellakopulos and P. Laubereau, *Angew. Chem., Intern. Ed. Engl.*, 4 (1965) 878.

13 F. Baumgärtner, E.O. Fischer, B. Kanellakopulos and P. Laubereau, *Angew. Chem., Intern. Ed. Engl.*, 5 (1966) 134.

14 R. Pappalardo, W.T. Carnall and P.R. Fields, *J. Chem. Phys.*, 51 (1969) 842.

15 F. Baumgärtner, E.O. Fischer, H. Billich, E. Dornberger, B. Kanellakopulos, W. Roth and L. Stieglitz, *J. Organomet. Chem.*, 22 (1970) C17.

16 P.G. Laubereau and J.H. Burns, *Inorg. Nucl. Chem. Letters*, 6 (1970) 59.

17 P.G. Laubereau and J.H. Burns, *Inorg. Chem.*, 9 (1970) 1091.

18 E.O. Fischer and A. Treiber, *Z. Naturforsch.*, 17b (1962) 276.

19 A.F. Reid and P.C. Wailes, *Inorg. Chem.*, 5 (1966) 1213.

20 F. Baumgärtner, E.O. Fischer, B. Kanellakopulos and P. Laubereau, *Angew. Chem.*, 81 (1969) 182.

21 E.O. Fischer and Y. Hristidu, *Z. Naturforsch.*, 17b (1962) 275.

22 F. Baumgärtner, E.O. Fischer, B. Kanellakopulos and P. Laubereau, *Angew. Chem.*, 80 (1968) 661.

23 P.G. Laubereau (according to ref. 7).

24 G.L. ter Haar and M. Dubeck, *Inorg. Chem.*, 3 (1964) 1648.

25 R.D. Fischer, R. von Ammon and B. Kanellakopulos, *J. Organomet. Chem.*, 25 (1970) 123.

26 E.O. Fischer, P. Laubereau, F. Baumgärtner and B. Kanellakopulos, *J. Organomet. Chem.*, 5 (1966) 583.

27 W.T. Carnall, P.R. Fields and R.G. Pappalardo, in M. Cais (Ed.), *Progress in Coordination Chemistry*, Elsevier, Amsterdam, 1968, p. 411.

28 C.-H. Wong, T.-Y. Lee and Y.-T. Lee, *Acta Cryst.*, 25B (1969) 2580.

29 R.D. Fischer, P. Laubereau and B. Kanellakopulos, Westdeutsche Chemiedozententagung, Hamburg (1968).

30 P.G. Laubereau, *Inorg. Nucl. Chem. Letters*, 6 (1970) 611.

31 B. Kanellakopulos, F. Baumgärtner and E. Dornberger, unpublished results (according to ref. 7).

32 B. Kanellakopulos, Proc. III Intern. Protactinium Conf., Schloss Elmau, Germany (1969). ZAED Report BMBW-FB K71-17, p. 24-1 (1971).

33 C.-H. Wong, T.-U. Yen and T.-Y. Lee, *Acta Cryst.*, 18 (1965) 340.

34 B. Kanellakopulos, E. Dornberger, R. von Ammon and R.D. Fischer, *Angew. Chem.*, 82 (1970) 956.

35 J. Müller, *Chem. Ber.*, 102 (1969) 152.

36 P. Zanella, S. Faleschini, L. Doretti and G. Faraglia, *J. Organomet. Chem.*, 26 (1971) 353.

37 Ethyl Corporation report TID-19367 (1963).

38 A. Streitwieser Jr. and M. Yoshida, *J. Am. Chem. Soc.*, 91 (1969) 7528.

39 A. Streitwieser Jr. and U. Müller-Westerhof, *J. Am. Chem. Soc.*, 90 (1968) 7364.

40 D.G. Karracker, J.A. Stone, E.R. Jones Jr. and N. Edelstein, *J. Am. Chem. Soc.*, 92 (1970) 4841.

41 A. Zalkin and K.N. Raymond, *J. Am. Chem. Soc.*, 91 (1969) 5667.

42 F. Mares, K. Hodgson and A. Streitwieser Jr., *J. Organomet. Chem.*, 24 (1970) C68.

43 E. Chauvenet, *Compt. Rend.*, 151 (1910) 387; *Ann. Chim. Phys.*, 23 (1911) 425.

44 O. Schmitz-Dumont and F. Raabe, *Z. Anorg. Allgem. Chem.*, 277 (1954) 297.

45 G. Tschirne and D. Naumann, *Z. Anorg. Allgem. Chem.*, 354 (1967) 50.

46 O. Schmitz-Dumont, F. Füchtenbusch and H. Schneiders, *Z. Anorg. Allgem. Chem.*, 277 (1954) 315.

47 G.W. Watt and S.C. Malhotra, *J. Inorg. Nucl. Chem.*, 11 (1959) 255.

48 H.J. Berthold and H. Knecht, *Angew. Chem.*, 77 (1965) 428; *Z. Anorg. Allgem. Chem.*, 356 (1968) 151.

49 J. Selbin, M. Schober and J.D. Ortego, *J. Inorg. Nucl. Chem.*, 28 (1966) 1385.

50 W. Burk and D. Naumann, *Z. Chem.*, 9 (1969) 189.

51 D.C. Bradley and M.H. Gitlitz, *J. Chem. Soc. (A)*, (1969) 980.

52 K.W. Bagnall and E. Yanir, unpublished observations.

53 R.G. Jones, G. Karmas, G.A. Martin Jr. and H. Gilman, *J. Am. Chem. Soc.*, 78 (1956) 4285.

54 F.A. Cotton and F. Zingales, *J. Am. Chem. Soc.*, 83 (1961) 351.

Chapter 12

THE f ORBITALS, AND THE MAGNETIC PROPERTIES
AND ABSORPTION SPECTRA OF THE ACTINIDES

1. INTRODUCTION

The behaviour of f electron systems and, in particular, the paramagnetism and absorption spectra of ions in which unpaired f electrons are present, has attracted an increasing interest in recent years. This has brought to light marked differences in the observed behaviour of isoelectronic ions of the lanthanide and actinide series. Although a detailed interpretation of the observed phenomena is, in many cases, still lacking, and the mathematical treatment required to handle the problem is far beyond the scope of this monograph, it is appropriate to include a brief account of these aspects at this point, without going deeply into the subjects under discussion.

The sections dealing with f orbitals and magnetic properties have been provided with brief theoretical introductions which, it is hoped, may be useful to readers, but in general, this short chapter is intended to bring together and into perspective much that has had passing mention throughout the book. Those seeking further or more detailed information will find several references here which may satisfy this requirement.

2. THE f ORBITALS

The expressions for the f orbital wave functions are similar in form to those for the s, p and d orbitals, the spherical harmonics being[1]:

$$Y_3^0 = (7/8)^{\frac{1}{2}} (5 \cos^3 \vartheta - 3 \cos\vartheta)(2\pi)^{-\frac{1}{2}}$$

$$Y_3^{\pm 1} = (21/32)^{\frac{1}{2}} (5 \cos^2 \vartheta - 1)\sin\vartheta (2\pi)^{-\frac{1}{2}} e^{\pm i\phi}$$

$$Y_3^{\pm 2} = (105/16)^{\frac{1}{2}} \sin^2 \vartheta \cos\vartheta (2\pi)^{-\frac{1}{2}} e^{\pm 2i\phi}$$

$$Y_3^{\pm 3} = (35/32)^{\frac{1}{2}} \sin^3 \vartheta (2\pi)^{-\frac{1}{2}} e^{\pm 3i\phi}$$

The functions $Y_3^{\pm m}$ include the expression $e^{\pm im\phi}$, where m takes the integral values 1, 2 or 3, and combinations of these functions are selected so as to eliminate i $(= \sqrt{-1})$. In this way the general set of real *f* orbitals, as distinct from the *imaginary* forms which include terms in i, are constructed; the results are given in Table 34.

TABLE 34

THE GENERAL SET OF *f* ORBITALS

Orbital[a]	Abbreviation	Wave function
$\frac{1}{2} z(5 z^2 - 3r^2)$	f_{z^3}	Y_3^0
$\frac{3}{2} x(5 z^2 - r^2)$	f_{xz^2}	$\frac{1}{2}(Y_3^1 + Y_3^{-1})$
$\frac{3}{2} y(5 z^2 - r^2)$	f_{yz^2}	$-\frac{1}{2} i(Y_3^1 - Y_3^{-1})$
$\frac{\sqrt{15}}{2} z(x^2 - y^2)$	$f_{z(x^2-y^2)}$	$\frac{1}{2}(Y_3^2 + Y_3^{-2})$
$\sqrt{15}\, xyz$	f_{xyz}	$-\frac{1}{2} i(Y_3^2 - Y_3^{-2})$
$\frac{\sqrt{15}}{2} x(x^2 - 3 y^2)$	$f_{x(x^2-3 y^2)}$	$\frac{1}{2}(Y_3^3 + Y_3^{-3})$
$\frac{\sqrt{15}}{2} y(3 x^2 - y^2)$	$f_{y(3 x^2-y^2)}$	$-\frac{1}{2} i(Y_3^3 - Y_3^{-3})$

[a]The polar forms are also given by Eisenstein[2].

The general set of *real f* orbitals does not contain any groups of orbitals which would be triply degenerate in a cubic ligand field. Since group theory requires that the seven *f* orbitals split in such a field $(O, O_h, T_d,$ etc.) into two groups of triply degenerate orbitals and a seventh non-degenerate orbital, it is necessary to derive a second basis set of *real f* orbitals, the cubic set, which satisfies these requirements. This cubic set is derived from the general set by combining four orbitals from that set in pairs to give four new ones, three of the general set $(f_{z^3}, f_{z(x^2-y^2)}$ and f_{xyz} being common to both sets. The cubic set, and the combinations which give rise to the individual orbitals, are given in Table 35. It should be noted that both sets of orbitals satisfy the

TABLE 35

THE CUBIC SET OF f ORBITALS

Designation	Orbital	Abbreviation	Combination[3]
f_δ, t_{1u}	$\begin{cases} \frac{1}{2}x(5\,x^2-3\,r^2) \\ \frac{1}{2}y(5\,y^2-3\,r^2) \\ \frac{1}{2}z(5\,z^2-3\,r^2) \end{cases}$	f_{x^3} f_{y^3} f_{z^3}	$-\frac{1}{4}[\sqrt{6}f_{xz^2}-\sqrt{10}f_{x(x^2-3\,y^2)}]$ $-\frac{1}{4}[\sqrt{6}f_{yz^2}+\sqrt{10}f_{y(3\,x^2-y^2)}]$
f_β, a_{2u}	$\sqrt{15}\,xyz$	f_{xyz}	
f_ϵ, t_{2u}	$\begin{cases} \frac{1}{2}\sqrt{15}[x(z^2-y^2)] \\ \frac{1}{2}\sqrt{15}[y(z^2-x^2)] \\ \frac{1}{2}\sqrt{15}[z(x^2-y^2)] \end{cases}$	$f_{x(z^2-y^2)}$ $f_{y(z^2-x^2)}$ $f_{z(x^2-y^2)}$	$\frac{1}{4}[\sqrt{10}f_{xz^2}+\sqrt{6}f_{x(x^2-3\,y^2)}]$ $\frac{1}{4}[\sqrt{10}f_{yz^2}-\sqrt{6}f_{y(3\,x^2-y^2)}]$

necessary mathematical criterion, namely that all orbitals in each set are orthogonal to all other orbitals in the same set.

The shapes and spatial orientations of the f orbitals are shown in Fig. 29; from these data it is then possible to draw up splitting diagrams for crystal fields of any desired symmetry. Two examples, redrawn from the literature[3], appear in Fig. 30 which shows the splitting for octahedral and hexagonal bipyramidal fields respectively. The result for octahedral coordination is very simple to deduce; the six ligands lie at the centres of the faces of a cube, the origin of axes being as shown for the f_{xyz} orbital in Fig. 29. It is then immediately obvious that the f_{x^3}, f_{y^3} and f_{z^3} (t_{1u}) orbitals, the lobes of which are directed towards the ligands, are of highest energy. The lobes of the t_{2u} set of orbitals, $f_{x(z^2-y^2)}, f_{y(z^2-x^2)}$ and $f_{z(x^2-y^2)}$, point towards the centres of the edges of the cube, and so must be of lower energy than orbitals of the t_{1u} set, but higher in energy than the f_{xyz} orbital, the lobes of which are directed towards the corners of the cube.

A set of character tables for all symmetry groups that are likely to be encountered in practice, and which include the irreducible representations to which the seven f orbitals belong, is now available[4]. Using such tables, one can apply standard group theoretical methods to establish which atomic orbitals, including f orbitals, may be combined to

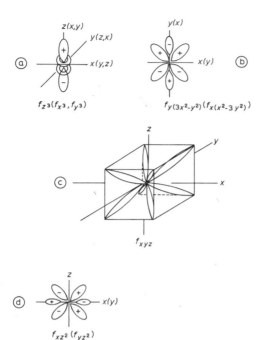

$f_{z^3}(f_{x^3}, f_{y^3})$ $f_{y(3x^2-y^2)}(f_{x(x^2-3y^2)})$

f_{xyz}

$f_{xz^2}(f_{yz^2})$

Fig. 29. The *real* forms of the angular dependence functions for *f* orbitals. (*b*) Six identical lobes; (*c*) eight identical lobes; $f_{z(x^2-y^2)}$, $f_{x(z^2-y^2)}$ and $f_{y(z^2-x^2)}$ are obtained by rotating the f_{xyz} orbital by $45°$ about the *z*, *x* and *y* axes respectively; (*d*) four large and two small lobes.

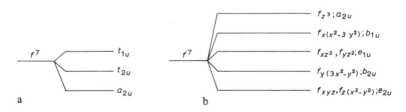

Fig. 30. The splitting of *f* orbitals in (*a*) an octahedral and (*b*) a hexagonal bipyramidal crystal field. Redrawn (not to scale) after figures in ref. 3, by permission of the American Chemical Society.

form bonding orbitals for various spatial arrangements of the bonds. Kettle and Smith have published[5] a compilation of the irreducible representations that are spanned by the seven *f* orbitals in the common point group symmetries, obtained by such a treatment. Their work shows that there are several point groups for which one or more irreducible representations are spanned by f orbitals exclusively. In other point group symmetries where the calculation indicates that there is a choice of p or f orbitals, it is probable that there is no f orbital involvement in the bonding, but in such cases there is no evidence one way or the other.

Examples of coordination geometries where f orbitals seem to be involved include cubic 8-coordination (O_h) and regular hexagonal bipyramidal 8-coordination (D_{6h}); examples of compounds which have been shown to have coordination geometries close to these are mentioned in earlier chapters. Kettle and Smith's paper[5] certainly provides a useful geometrical test of possible f orbital involvement in bonding, but treatments of this kind do not, of course, prove that f orbitals are actually involved.

Their preliminary treatment[5] has been extended in a subsequent paper[6]; in this the possible steric consequences arising from the presence of f electrons on the metal atom are considered. The conclusion reached was that an unsymmetrical f orbital occupation could have steric consequences, although these would be small, whereas in the case of weak interactions involving symmetrically occupied f orbitals with the orbitals of the surrounding ligands, there would be no directional effect.

This is clearly a field in which more work needs to be done, preferably in conjunction with full structure determinations of a really wide variety of actinide compounds, for, until more structural information is available, it is unlikely that major progress will be made.

3. MAGNETIC PROPERTIES OF THE ACTINIDE IONS

The magnetic properties of the actinide ions are best discussed in relation to those of the lanthanide ions, since the behaviour of the ions of both series of elements differs markedly from that observed for the d transition element ions. A summary of the behaviour of the two f transition series ions is included in a recent review of magnetochemistry[7]

and also in a recent book on ligand-field theory[1]; in these fuller details of
the magnetic properties of individual compounds may be found.

The effective magnetic moments, observed at room temperature,
of most of the lanthanide ions agree well with those calculated from the
expression:

$$\mu_{eff} = g\,[J(J+1)]^{\frac{1}{2}}$$

This expression implies that the effective magnetic moment is
independent of temperature and that all compounds of the same
electronic configuration will have the same effective magnetic moment.

For the first condition to be so, the ground state must be well
separated ($\sim$ 2000 cm^{-1}, as in the Pr^{3+} ion) from the next higher-lying
state; this implies that there should be no thermal population of the
latter. However, the requirement fails for the lanthanide f^5 and f^6 ions,
in which the ground states are $J = 5/2$(Sm^{3+}) and $J = 0$(Sm^{2+}, Eu^{3+});
ions of both of these configurations have quite large, temperature-
dependent moments arising from thermal population of states that are
lower-lying than is the case with the other lanthanide ions[8]. These low-
lying states associated with f^5 and f^6 ions are $J = 7/2$ (f^5, Sm^{3+}) and
$J = 1$ (f^6, Sm^{2+}, Eu^{3+}).

The observed room-temperature magnetic moments of the
lanthanide ions are almost, but not quite, independent of the magnetic
dilution of the compound studied. Until recently it had been generally
assumed that the moments were completely independent of the nature
of the compound, and that the lanthanide ions could be considered as
gas-phase ions, because of the shielding of the $4f$ electrons from the
influence of other atoms by the overlying $5s$ and $5p$ electrons. This, in
turn, could be ascribed to the contracted state of the $4f$ orbitals arising
from the rather high effective nuclear charge to which they are exposed
compared to that experienced by d electrons in the d transition series
and $5f$ electrons in the actinides.

However, even the lanthanide case is not quite so straightforward
as this; with the exception of the f^7 ions (Eu^{2+}, Gd^{3+}), the Weiss

constants are not zero, and the effective magnetic moments are not independent of temperature, so that the agreement of the calculated and observed room temperature magnetic moments is rather fortuitous. In the $4f$ series spin-orbit coupling is a relatively minor perturbation compared to inter-electronic repulsions, the values of the former lying between 600 and 2500 cm^{-1}, but they are appreciably greater than the ligand-field splittings, which are probably about 1% of those observed for the d transition metal ions[7], about 100 cm^{-1}, that is, of the same order as kT. The ligand-field splittings are probably of a similar order in the lanthanide-like transplutonium elements, but it is difficult to assess the magnitude of this effect for the f^1, f^2 and f^3 actinide ions and one can only say that it is probable that spin-orbit coupling in these actinide ions, at any rate in their higher oxidation states, is probably of the same order as the ligand field splittings. That is, the ligand-field splittings are much larger for actinide ions than for lanthanide ions.

Even for lanthanide ions, however, dilution experiments have shown[9] that magnetic concentration has a slight, but measurable effect, which could be explained in terms of the perturbation of the electronic states of the Tm^{3+} and Yb^{3+} ions, investigated in this study, by the crystal field. Although this ligand-field splitting is usually disregarded as being negligible compared with the separation of adjacent states, each state can be split by about 100 cm^{-1} and the ground states are split by about kT(210 cm^{-1} at room temperature), from which the temperature dependence of the effective magnetic moments arises.

Such effects are even more marked in the actinide series, where the temperature dependence of the effective magnetic moments is much greater than in the $4f$ series; this is likely to be due largely to thermal population of low lying states coupled with the greater influence of the ligand field on the $5f$, than on the $4f$, ions.

In both of the f transition series it is difficult to evaluate the magnitudes of the interactions of f orbitals with the ligand orbitals because of the complex shape of the f orbitals and the low symmetry of the environment of the metal ions in so many of their complexes. Single crystal studies of certain of the more highly symmetrical lanthanide species have been reported[10], but the mathematical treatment is more

specialised than can be dealt with here. It certainly appears that every
lanthanide and actinide compound will require individual examination,
and many such examinations will be needed before a comprehensive
picture can be expected to appear.

The available data have been summarised[7] and other results are to
be found in papers describing the preparation of complexes of the
actinide halides, nitrates and, in some cases, organometallic compounds.
It must be stressed, however, that there are very few detailed inter-
pretations of the magnetic behaviour of actinide ions. There is a superficial, qualitative similarity in the magnetic behaviour of isoelectronic
$4f$ and $5f$ ions (Fig. 31), but even here the differences (notably between
Eu^{3+} and Am^{3+}) are rather obvious, and the moments plotted are all
room temperature values. In addition, the actinide ions used as
examples in Fig. 31 are not all in the same oxidation state.

4. THE ABSORPTION SPECTRA OF ACTINIDE IONS

A review[11] which has recently appeared deals in detail with the
individual f^n configurations and the effects of the ligands on the
transition energies and the intensities of the individual transitions, as
well as with the use of the spectra for identification purposes. It is re-
commended as a supplement to the brief resumé that follows.

Three main types of electronic transition are observed in the
spectra of lanthanide and actinide ions. The first group consists of the
relatively weak, extremely narrow bands that stem from the Laporte-
forbidden $f-f$ transitions. These are easily identifiable in the spectra
because of their sharpness. The second group consists of the intense,
quite broad bands that arise from the allowed $nf^q \rightarrow nf^{q-1} (n + 1)d$
transitions; these often, but not always, appear in the ultraviolet region.
Finally there are the intense, broad charge-transfer bands that arise
from the transfer of an electron from the ligand to a metal orbital.
These usually appear in the ultraviolet, but may also appear in the near
u.v. region of the visible spectrum, and give rise to the intense colours
observed for many actinide ions.

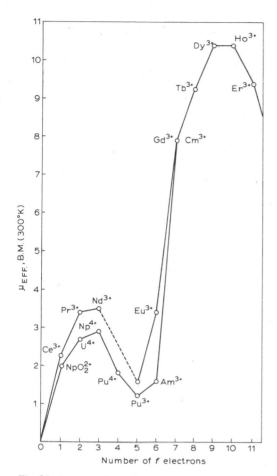

Fig. 31. A comparison of the room temperature magnetic moments of some lanthanide and actinide ions (data from ref. 7).

The spectra arising from internal f–f transitions are rather complex because of the large numbers of possible states of the f^n configurations, and the complexity is further increased by the splitting of these states as a result of spin-orbit interaction and ligand-field effects. The magnitude of the latter effect, as mentioned earlier, is small in the case of the lanthanides and the spectrum of an individual lanthanide does not differ very markedly from one compound to another. The splitting

References p. 262

of the free ion J levels for the tervalent actinides is likewise small, being
about twice that for the lanthanides, but in the higher oxidation states
of the earlier actinides the splitting is of the same magnitude as that
caused by spin-orbit interaction. This last, for the U^{4+} ion in an
octahedral field, is more than twice that for its isoelectronic lanthanide
analogue[12], Pr^{3+}.

The simplest case to consider is the quite easily accessible $5f^1$
system, in the shape of the octahedral uranium(V) anions UX_6^- (X = F,
Cl, Br), which has been studied by Ryan[13]. The sharpness of the bands
arising from the f-f transitions is evident from the spectra shown in Fig.
32. The two energy levels are $^2F_{5/2}$ (six-fold degenerate) and $^2F_{7/2}$
(eight-fold degenerate) and there are no interelectronic repulsions to be

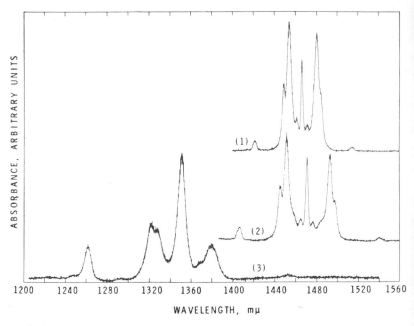

Fig. 32. The absorption spectra of uranium(V) hexahalocomplexes in the
1200–1560 mμ region: (1) solid $(C_2H_5)_4NUBr_6$, (2) solid $(C_2H_5)_4NUCl_6$ and
(3) solid $(C_6H_5)_4AsUF_6$. (J.L. Ryan, $J.$ $Inorg.$ $Nucl.$ $Chem.$, 33 (1971) 153).
Reprinted by permission of Pergamon, Oxford.

considered in the single-electron system. However, the spectra of these octahedral f^1 anions are made quite complex because of the vibronic coupling which arises from electron-vibration interaction. If these anions were of perfect octahedral symmetry, purely electronic bands would not be observed in their spectra, for these would be forbidden as electric-dipole transitions since the metal ion lies at a centre of inversion. Consequently only the vibronic transitions should appear. In fact, even a very slight distortion of the octahedron increases the intensities of the purely electronic transitions in such complexes to an extent sufficient for them to be observed[14-16]. These aspects are discussed in some detail by Ryan[13].

The same situation arises with octahedral f^2 complex ions, UX_6^{2-}, the spectra of which have been analysed[12, 14]. Here the octahedron is distorted to some extent when the cation is such that it can become involved in hydrogen bonding to the anion[16]; examples of cations for which this can occur are the mono-, di- and tri-alkylammonium cations, but not, of course, the tetraalkylammonium cation. The marked difference between the spectra of the UBr_6^{2-} anion in a compound of this type and one where no hydrogen bonding can occur is shown in Fig. 33.

The internal f-f transitions also show a quite large nephelauxetic effect; thus with the UX_6^{2-} anions, the spectrum of the UBr_6^{2-} ion is shifted by 3–5% towards lower frequency, as compared with that of the UCl_6^{2-} ion, and that of the UI_6^{2-} ion is shifted by a further 3–5% in the same direction[17]. The spectra of the three ions are shown in Fig. 34.

In general, it appears that the f-f transition spectra of the actinide ions are more sensitive to the environment of the metal ion than is the case with the lanthanides, where the main features of the spectra remain almost identical irrespective of the environment. Indeed, the spectrum of a given actinide ion may become almost unrecognisable when the environment is changed drastically. This is demonstrated by the spectra of the two uranium(IV) compounds, $UCl_4 \cdot 2$ TPPO and $U(NCS)_4 \cdot 4$ TPPO (TPPO = $(C_6H_5)_3PO$) shown in Fig. 35.

The difficulty in arriving at a complete interpretation makes it too difficult to use the u.v./visible spectrum of an actinide complex as a means of deducing the symmetry of the environment of the metal ion.

References p. 262

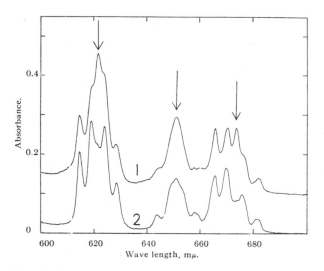

Fig. 33. The effect of hydrogen bonding on the absorption spectrum of UBr_6^{2-}. $1.7 \cdot 10^{-2}$ M $[(C_2H_5)_4N]_2UBr_6$ in (1) 1 M $(C_2H_5)_2NH_2ClO_4$ in nitromethane (absorbance scale displaced 0.1) and (2) in nitromethane (J.L. Ryan, *Inorg. Chem.*, 3 (1964) 211). Copyright (1964) by the American Chemical Society and reprinted by permission of the copyright holder.

However, such spectra are quite valuable in this sense when they are used as a "fingerprint" technique, for if the spectra of a given ion in a number of known environments are available, then a comparison of the spectrum of the species for which the symmetry is unknown with these reference spectra will often yield useful information concerning the symmetry of the former. For instance, by the use of this type of technique, it has been shown that an apparently quinquevalent uranium species, a complex with N,N-dimethylacetamide(DMA), $UOBr_3 \cdot 2.5$ DMA, really consists of a uranyl(VI) complex cation and a uranium(IV) anion[18], $[UO_2(DMA)_5]^{2+}[UBr_6]^{2-}$, whereas in the analogous uranium(IV) complex, $UCl_4 \cdot 2.5$ DMA, both octahedral and low symmetry uranium(IV) sites are present[19]. Applications of this kind are particularly useful in the identification of the species present in molten salts and in the liquid phases of solvent-extraction systems[11].

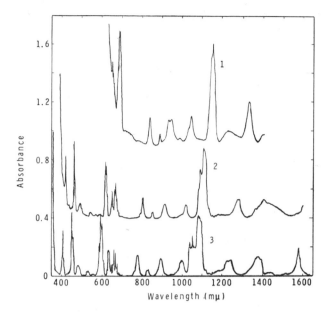

Fig. 34. The visible and near infrared absorption spectra of acetonitrile solutions of UX_6^{2-}. (1) $5.6 \cdot 10^{-2} M$ UI_6^{2-} (U(IV) iodide in concentrated tributylpropyl-ammonium iodide) in a 1.00-cm cell (absorbance scale displaced 0.9), (2) $2.4 \cdot 10^{-2} M$ $[(C_2H_5)_4N]_2 UBr_6$ in a 2.00-cm cell (absorbance scale displaced 0.4) and (3) $3.6 \cdot 10^{-2} M$ $[(C_2H_5)_4N]_2 UCl_6$ in a 2.00-cm cell (J.L. Ryan and C.K. Jørgensen, *Mol. Phys.*, 7 (1963) 17). Reproduced by permission of the publisher, Taylor and Francis, London.

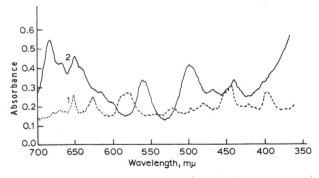

Fig. 35. Solid reflectance spectra of (1) 6-coordinate uranium(IV) in $UCl_4 \cdot 2$-$(C_6H_5)_3PO$ (dotted line) and (2) 8-coordinate uranium(IV) in $U(NCS)_4 \cdot 4$-$(C_6H_5)_3PO$ (continuous line).

References p. 262

5. REFERENCES

1 B.N. Figgis, *Introduction to Ligand Fields*, Interscience, New York, 1966, p. 8.
2 J.C. Eisenstein, *J. Chem. Phys.*, 25 (1956) 142.
3 H.G. Friedman, G.R. Choppin and D.G. Feuerbach, *J. Chem. Educ.*, 41 (1964) 354.
4 J.A. Salthouse and M.J. Ware, *Group Character Tables and Related Data*, Cambridge University Press, 1972.
5 S.F.A. Kettle and A.J. Smith, *J. Chem. Soc. (A)*, (1967) 688.
6 S.F.A. Kettle and A.J.P. Pioli, *J. Chem. Soc. (A)*, (1968) 122.
7 B.N. Figgis and J. Lewis in H.B. Jonassen and A. Weissberger (Eds.), *Technique of Inorganic Chemistry*, Vol. 4, Interscience, New York, 1965, p. 137.
8 J.H. van Vleck, *Theory of Electric and Magnetic Susceptibilities*, Oxford University Press, 1932.
9 L.M. Engelhardt and B.N. Figgis, *J. Chem. Soc. (A)*, (1970) 415.
10 M. Gerloch and D.J. Mackey, *J. Chem. Soc. (A)*, (1970) 3030, 3040; (1971), 2605, 2612.
11 J.L. Ryan, *Intern. Rev. Sci.*, 7 (1972) 323; Med. Techn. Publ., Aylesbury and Butterworth, London.
12 R.A. Satten, D.J. Young and D.M. Gruen, *J. Chem. Phys.*, 33 (1960) 1140.
13 J.L. Ryan, *J. Inorg. Nucl. Chem.*, 33 (1971) 153.
14 S.A. Pollack and R.A. Satten, *J. Chem. Phys.*, 36 (1962) 804.
15 J.R. Morrey, *Inorg. Chem.*, 2 (1963) 163.
16 J.L. Ryan, *Inorg. Chem.*, 3 (1964) 211.
17 J.L. Ryan and C.K. Jørgensen, *Mol. Phys.*, 7 (1963) 17.
18 F. Lux, G. Wirth and K.W. Bagnall, *Chem. Ber.*, 103 (1970) 2807.
19 K.W. Bagnall, unpublished observations.

Subject Index